The Fidelity of Isopleth Maps

AN EXPERIMENTAL STUDY

BY

Mei-Ling Hsu

AND

Arthur H. Robinson

UNIVERSITY OF MINNESOTA PRESS

MINNEAPOLIS

© Copyright 1970 by the University of Minnesota. All rights reserved. Printed in the United States of America at the North Central Publishing Company, St. Paul. Published in Great Britain, India, and Pakistan by the Oxford University Press, London, Bombay, and Karachi, and in Canada by the Copp Clark Publishing Co. Limited, Toronto

Library of Congress Catalog Card Number: 71-99127

Preface

ONE of the challenging problems in the field of cartography is to represent a three-dimensional distribution on a two-dimensional map. Over the years a number of techniques have been developed to depict the real or the abstract third dimension on the flat map, techniques ranging from such a simple form as marking spot elevations on the map to the graphically sophisticated method of plastic shading and the conceptually sophisticated processes used in choropleth, dasymetric, and isarithmic maps. Each has its own advantages and limitations, and all include some amount of error introduced by the mapping process itself. Relatively little attention, however, has been devoted by cartographers to rigorous analysis of these errors and, consequently, of the fidelity of mapping systems. The isarithmic method is no exception. This study is devoted to broadening our understanding of one class of isarithms, the isopleth.

The Graduate School of the University of Wisconsin through its Research Committee has generously supported a continuing program of research into the fundamental characteristics of the isarithmic map. This is the first major report of some of the results. The support provided not only research assistantships, but a considerable amount of student help in the enormously tedious and time-consuming operations attendant on this sort of experimental study. Literally thousands of measurements by planimeter, and their tallying, checking, and entering, were required to derive the needed data. The Graduate School also provided the necessary funds for the considerable amount of automatic data processing in the University of Wisconsin Computing Center.

The authors are grateful to the Cartographic Research Laboratory of the Department of Geography at the University of Minnesota and the Cartographic Laboratory at the University of Wisconsin for their expertise and assistance in connection with the preparation of the illustrations. We also wish to express gratitude to our departmental colleagues at these universities, who have contributed much in many discussions. Especial thanks are due to Professor Philip Porter of the Department of Geography at the University of Minnesota and to Professor Norman R. Draper of the Department of Statistics at the University of Wisconsin. Needless to say, any deficiencies in the study and in the handling of the data are solely the responsibility of the authors.

M.-L. H.
A. H. R.

Minneapolis and Madison

Table of Contents

1. QUANTITATIVE MAPPING AND THE ISARITHM   3

The Isarithmic Technique in General, 4. Isopleth Mapping, 5. Cartographic Generalization and the Isarithmic Interval, 6. Error Characteristics of the Isopleth Map, 8. The Experimental Design, 13.

2. THE PREPARATION OF THE EXPERIMENTAL DATA   14

Selection of the Original Distributions, 14. Selection of Unit Area Patterns, 19. Derivation of Unit Area Data, 19. Transformation to a Hexagonal Pattern, 24. The Resultant Isopleth Maps, 27.

3. TESTING AND SAMPLING PROCEDURES   28

Evaluation of Isarithmic Accuracy, 28. Sampling Procedure, 29. Level of Generalization, 50.

4. STATISTICAL ANALYSES OF THE FIDELITY OF THE ISOPLETH MAPS   53

The Analysis of Variance, 53. Magnitudes of Map Discrepancies, 55. Dispersion of Discrepancies, 57. Algebraic Values of Discrepancies, 61.

5. VISUAL ANALYSIS OF THE FIDELITY OF THE ISOPLETH MAPS   64

Effect of the Unit Area Pattern, 64. Effect of the Hexagon Size, 66. Effect of the Surface Configuration, 67. Relative Significance of the Variables, 69.

6. GENERAL CONCLUSIONS   72

NOTES, 77.   BIBLIOGRAPHY, 83.   INDEX, 89.

Plates 1 to 16 appear between pages 33 and 48

The Fidelity of Isopleth Maps

1

Quantitative Mapping and the Isarithm

THE cartographic portrayal of spatial arrays of numbers is commonplace today. The average citizen sees many such representations, ranging from those depicting voting patterns in national elections to the familiar topographic maps which show the distribution of elevations as measured from some arbitrary datum surface. At the present time maps depicting numerical distributions are increasingly being produced by machine methods through the use of the computer and various kinds of stereo devices. It is to be expected that their production, by hand and by machine, will increase, not only because they are easier to make but because there are more numbers than ever before.

The mapping of numbers naturally had to wait for the numbers to be available. The first isarithmic maps were prepared for navigators who needed to know the depths of water in the relatively shallow rivers, harbors, and coastal areas where they sailed. The oldest known example is a manuscript map of the channel of the River Spaarne, drawn in 1584 by the Dutch surveyor Pieter Bruinss, on which soundings were shown by lines of equal depth (isobaths) below sea level.[1] From this beginning, there was an increasing use of the technique by such men as Pierre Ancelin (1697), Luigi de Marsigli (1725), Nicolas Cruquius (1730), and Philippe Buache (1737).[2] For some time, however, the technique was limited to the representation of the under-water surface, until, as Dainville so well puts it, the line of equal elevation above sea level (isohypse or, commonly, contour) of the topographic maps of the land rose like Aphrodite out of the sea, and the technique of using lines of equal depth was transferred to the land. One of the first proposals (Milet de Mureau, 1749) for the use of contours on the land actually involved an arbitrary elevated plane of reference or datum from which "depths" were reckoned to the land below it. In the latter half of the eighteenth century — through the work of Jean Baptiste Marie Meusnier (1777), Marcellin du Carla (1782), Jean Louis Dupain-Triel (1791), and François N. B. Haxo (1801) — and during the early years of the nineteenth century, the contour on the land came of age, so to speak, and became a standard symbol.[3]

The mapping of another class of numbers also had its origin prior to the eighteenth century, and it too was founded on the needs of mariners. Over the earth the needle of the magnetic compass aligns itself with north-south at relatively few places. The departure, called declination and stated in degrees, is rather easily observed. Knowledge of the distribution of declination would clearly be of great use to navigators, both as an aid in setting bearings and, hopefully, as a help in determining longitude. The earliest such map is said to have been one made by the Jesuit Father Christoforo Borri of Milan (ca. 1630).[4] As described by Athanasius Kircher, the map employed lines of equal amounts of declination (isogones) to show the distribution, and this puts the symbolism in the same class as that of the isobath and the isohypse. In 1701 Edmund Halley published his famous isogonic chart of the Atlantic Ocean and a year later another chart that covered much of the world.[5] Halley's charts were a great success and in revised form were periodically issued well into the latter part of the eighteenth century. Other lines of equal value were used by students of magnetism during the eighteenth century, and by the beginning of the nineteenth century maps showing the distribution of various kinds of numbers related to earth magnetism were common.

The representation of a whole new class of spatially arrayed numbers began with the appearance in 1817 of Alexander von Humboldt's isothermal chart of temperatures.[6] During the succeeding thirty years the line as a symbol joining points of equal numerical value was applied to numerous phenomena of physical geography such as rainfall, air pressure, and thunderstorm frequency. Shortly after the middle of the nineteenth century this technique was expanded to include social and economic data such as population density and agricultural production.[7] Although the iso-terms by which generic classes of these various lines are now known did not come into use until later, the kinds of maps showing numerical distributions steadily increased in number and importance.[8]

Lines of equal value are not the only way numerical distributions may be represented. The techniques known as choropleth and dasymetric mapping are also used and they too had their origins early in the nineteenth century. The choropleth map, in which enumeration units such as states or counties are colored or shaded to indicate average values within each unit, was first published by Charles Dupin in 1827.[9] Similar maps of social statistics followed rapidly, and choropleth maps were common by the middle of the century. Dasymetric mapping, wherein shading or coloring is applied to "natural groupings," that is, class limits made independently of the enumeration boundaries, also had its beginning in the first half of the nineteenth century.[10]

Whatever the symbolism employed, the objective of the cartographer is to portray the distribution of numbers as efficiently as possible, and in order to do so he must first decide what characteristics of the spatial array are paramount. The primary interest may simply be the location of particular values, and by making the map he is merely providing the geographical dimension for what would otherwise be a less revealing tabular listing. Such an objective usually results in choropleth and dasymetric maps, which are generally restricted to social and economic data. When the primary concern is to symbolize the relation among the numbers so as to indicate the gradients, troughs, rises, and other configurational characteristics of what has been called the statistical surface, then lines of equal value are ordinarily employed.[11] The generic term for these kinds of lines is *isarithm*, and the present study focuses in general on the ability of the isarithmic technique to portray reality and more especially on the capabilities of one particular class of isarithms.

The Isarithmic Technique in General

As we have suggested, there are two classes of isarithms, exemplified by contours of the land on the one hand and by lines of equal population density on the other; the one class is separated from the other by the significant characteristic of the precision with which the lines can be located. Those, like contours, which can be precisely located on the map base, at least in theory, are called *isometric lines*, while the lines of the other class are called *isopleths*. Isometric lines, such as contours, the isotherms and isobars of the weather map, or lines showing the average percentages of rainy days, are so called because the values upon which they are based can actually be assigned to point locations on a map. In contrast isopleths, such as lines showing density of population or average land values, are distinguished by the fact that the point values upon which they are based are indeterminate in location because a two-dimensional area is involved in the derivation of the value.[12] In the following discussion, whenever statements are applicable to both classes of lines the generic term isarithm will be employed.

The isarithmic technique is used to portray the surface of a volume distribution[13] conceptually as opposed to a pictorial representation by, for example, plastic shading.[14] The statistical surface may be defined as consisting of z values varying in vertical dimension over the xy dimensions of the map base, thus forming a volume. Such a volume distribution must be assumed to form a continuous statistical surface; that is, any profile across the surface will be a continuous curve and not segments of straight lines. In reality, we can know completely only one such statistical surface, namely, the configuration of the land surface. This knowledge is possible because of photogrammetric methods, but for all other surfaces which we map, we must derive the continuous surfaces by inductive processes from limited, and sometimes isolated, sampling knowledge. For example, one must employ the records of average July temperatures from a limited group of weather stations in order to form the undulating surface of this temperature abstraction with reference to the defined datum, for instance, the zero degree of centigrade. An even greater abstraction is employed in mapping a distribution of population density.

With the assumption of a continuous statistical surface, the cartographer proceeds with his isarithmic mapping. At this point he must decide upon the interval for the isarithms (for instance, a uniform interval). A number of assumed planes, parallel to the adopted datum and spaced according to the distance of the isarithmic interval, are then made to intersect the imaginary statistical surface. Finally the traces of the intersections are orthogonally projected onto the datum and mapped as continuous curves. These are the isarithms; thus an isarithm is any trace of the intersection of a horizontal plane with a statistical surface. Each isarithm "passes through" all points which have the same height on the statistical surface. Therefore, the isarithm is often defined as being a curve of equal (height) values or a line which connects all points of a particular value (height). While these latter definitions are correct, they are not comprehensive.

A knowledge of the three-dimensional concept upon which most isarithmic mapping is founded is not only fundamental to the interpretation of the map, but also indispensable for analyzing the characteristics of the technique. This is evident from the fact that most authors construct some kind of diagram of three-dimensional appearance to supplement their discussion of the characteristics of the isarithmic map.

The commensurability of the isarithmic map is a valuable quality for mapping statistical surfaces. Since each isarithm represents a certain z value, the point values which lie between any pair of isarithms can be obtained by interpolation. Thus, theoretically, the z values of all points on an isarithmic map are known. This capability is very useful if further numerical analysis of the surface is desired or if the relation between two surfaces is to be studied. The accuracy of the interpolated values, and consequently the accuracy of any subsequent analyses, is affected, however, by a great many elements of the isarithmic mapping process. Some of these elements are treated specifically in this book. Generally speaking, however, except in the case of large-scale, modern topographic maps, interpolated values on an isarithmic map can provide no more than general indications of the z values at the concerned points. One may be able to claim that on a high quality topographic map the error involved in an interpolated value is less than one-half of the contour interval, but the same statement cannot be applied safely to maps of smaller scale. In short, the commensurability of the isarithmic method is an invaluable quality of large-scale maps of the land surface, but it must be considerably qualified with respect to small-scale maps.

Although we stressed previously that the isarithmic technique depicts a volume distribution conceptually rather than pictorially, an isarithmic map does have some pictorial quality. For instance, there is an approximation of vertically lighted shading as a consequence of the increase of the density of lines with increasing slope. In most cases, however, the pictorial quality of the map is a passive factor; that is, it is there if the reader is trained to see it. An experienced reader who is familiar with the conceptual assumptions of the technique will find it quite easy to visualize the form of the surface as shown by the characteristic shapes and patterns of spacings of the isarithms. In contrast to the quality of commensurability, the visual effectiveness of an isarithmic map is not affected so much by the scale and precision of the map as by the training and experience of the reader.

Isopleth Mapping

Consider now an isopleth map. Its preparation begins with the location of numbers (values) at specific places (control points) on a map. The control point values are assumed to occur as spot heights at various elevations, in proportion to their magnitudes, in the z dimension perpendicular to the plane of the map datum. Thus a control point value of 30 is twice as "high" as a value of 15. The set of control point values therefore establishes, in theory, a three-dimensional surface exactly as would numerical values of land elevation at those x and y control point locations. By interpolation among the control point values, isopleths are located to show the configuration of the "surface" defined by the control point values. In theory, then, an isopleth is like a contour in that it is a line of constant "elevation." By its inflections, spacings, and orientations, a set of isopleths, like a set of contours, portrays the various gradients, the directions of slopes, and the arrangements of ridges, troughs, rises, and hollows of an assumed three-dimensional surface. One reads an isopleth map precisely the way one does a contour map. But there is a vast difference between a contour map and an isopleth map.

The land surface exists. By field survey and photogrammetric methods one can ascertain the various z elevations of the land surface above a given horizontal datum, and basically the task facing the maker of a contour map of the land is simply to put the contours where they actually belong. This can be done within the limitations imposed by the scale and the desired accuracy.

The isopleth map is quite different. The configuration of the surface it delineates is a geographical abstraction — for example, the variations from place to place of the density of population, that is, the number of persons per square kilometer or mile. There is no limit to the number of such ratios or to the degree of their abstractness: they range from unusual topics like the percentage of a given blood type or the average IQ of schoolchildren to such common concepts as the percentage of farmland in cultivation or the average income per family. Because these data are ratios which either involve area in their definition or are derived from data enumerated by areal units, it is quite impossible to ascertain exactly where isopleths should be located.[15]

The numerical data to be mapped by isopleths always consist of percentages, ratios, proportions, and the like, derived from data gathered by enumeration districts (called unit areas) such as census tracts, counties, townships, parishes, boroughs, or the cells of any stratified system of enumeration used for geographical research. Consequently, each control point value, from the set of which one infers the statistical surface, is always an average of the magnitudes that occur in variable fashion throughout the given unit area.

An example from population distribution will clarify the problem. Although there may be more people living in the northwest half of a county than in the southeast, the single control point value for that county, on which the isopleth portrayal will depend, is obtained by dividing the total population of the county by its area. Within the northwest half there may also occur variations: a village may produce a peak of high density on the statistical surface and a strip of poorly drained land may occasion a trough of low density. These together with other kinds of differences can easily result in an undulating surface that may be quite complex. Yet in order to map the configuration of an abstract ratio surface by isopleth methods, these undulations of varying complexity must be submerged in spatial averages that apply to unit areas of various geographical areas and arrangements. Thus the population of a section of a state may be enumerated in the first instance by small census divisions, generally called towns, townships, parishes, or in the United States simply "minor civil divisions." The density per unit area for each small unit is calculated by dividing the population by the area, and these density quantities can then be mapped. At another level of generalization the populations of the townships can be aggregated by the counties of which they are parts, and a similar map can be prepared on the basis of the county data. Both maps will represent the same basic distribution but their appearance will be quite different, as Figure 1 shows. At an even higher level of generalization, one could proceed by state units.

The differences among the maps just described, covering the same area, arise from several causes, of which the most basic is the level of generalization, that is, the amount of detail shown. This factor, although it is the most fundamental one, is chosen by the cartographer and is generally within his control. Far more complex and technically more important from the cartographic point of view is the potential effect on the positions of the isopleths of such other factors as the comparative sizes and shapes of the enumeration districts employed in aggregating the data, the nature and complexity of the "real" distribution itself, and various other "error" effects of the procedures employed in the mapping process. Under some circumstances these factors may markedly change the directions and magnitudes of the gradients and even the elevations of the surface. In view of the number of variables in the isopleth mapping process, a cartographic question of fundamental importance arises: with what fidelity does an isopleth map actually portray the "true" configuration of the abstract distribution it represents and what are the error effects of the major variables in the mapping process.

Cartographic Generalization and the Isarithmic Interval

There are many aspects of generalization in isarithmic mapping; some, like the map scale, are common to all mapping techniques, while others are unique — for example, the smoothing of isarithms and the characterization of the overall nature of the continuous surface. The latter aspect is not only unique but fundamental in the sense that it occurs at the very beginning of the isarithmic mapping process.

One of the inherent limitations of the isarithmic technique is that a distribution can be represented by only a limited number of isarithms; that is, the distribution cannot be depicted totally. This follows from the assumption that the distribution forms a continuous undulating surface and that complete description would require the employment of an infinite number of isarithms. Since this is impossible, the cartographer must characterize the total distribution with a selected number of lines; accordingly the distribution is generalized in the sense that irregularities in the variations of the distribution which happen to occur within the dimen-

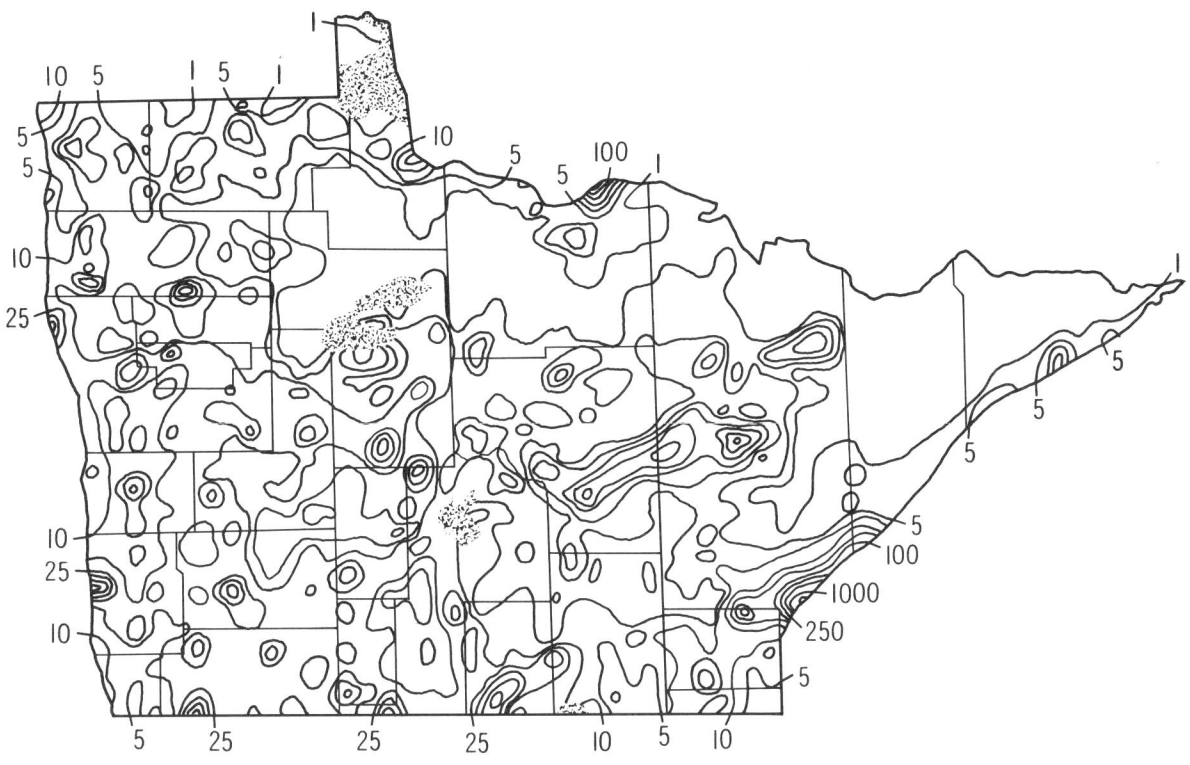
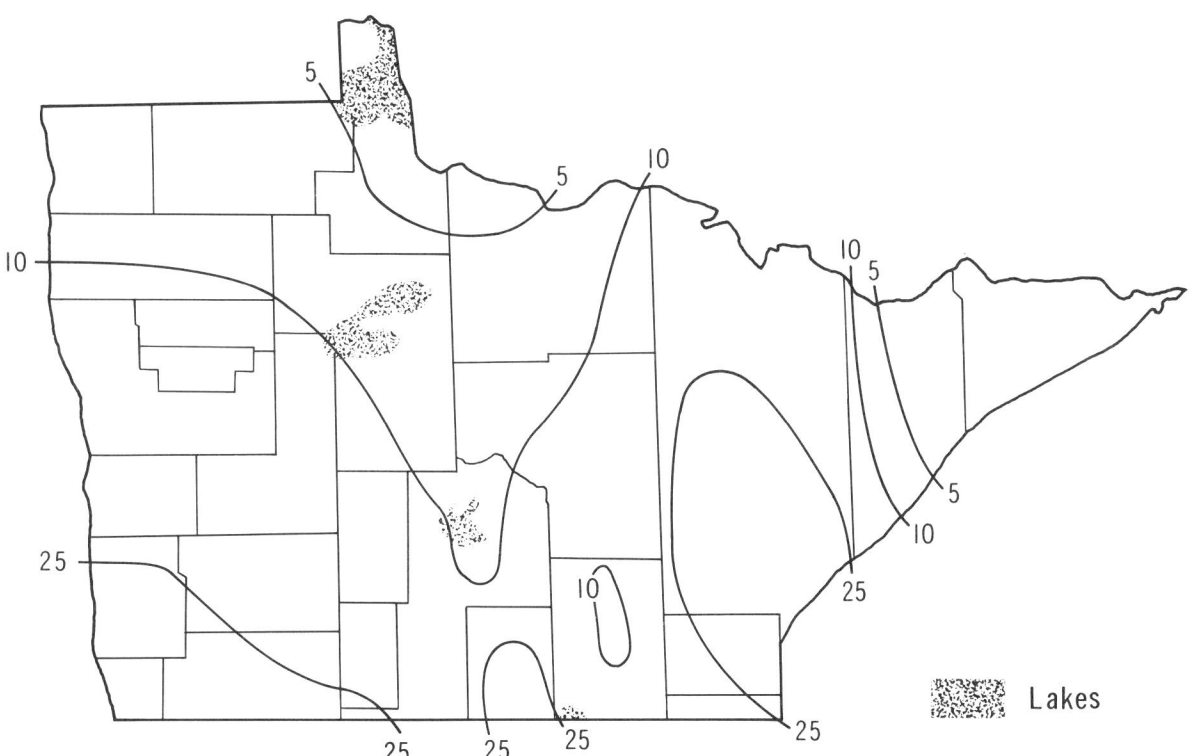

Figure 1. Two Isopleth Maps Showing Population Distribution in Northern Minnesota in 1950. The upper map, by John W. Webb, is based upon minor civil division enumeration units, the lower upon counties.

sional zone between any pair of isarithms are lost from the map.

The cartographic characterization of a continuous statistical surface is accomplished (1) by choosing a system for the isarithmic interval and (2) by selecting one from the infinite numbers of sets of isarithms specified by the interval system to represent the distribution. The choice of interval obviously affects the precision of a map, for the details of the distribution which lie between pairs of isarithms and within the interval are buried, so to speak. Therefore the larger the interval, the more detail will be lost. This aspect of generalization is not an error characteristic, but a limitation.[16] It is inherent in representation by isarithms, and the choice of an interval must be judged in terms of whether it is appropriate or inappropriate (for the purpose of the map), rather than right or wrong.

Ideally the interval should be small enough so that the lost information is kept to a minimum; in practice, however, the size of the interval must be consistent with the quality and quantity of the data. Since the choice of interval is an important aspect of isarithmic mapping, much work has been done with respect to the types of intervals in use, the method of selecting intervals, and so forth.[17] This work will not be reviewed here.

It should be stressed that several maps drawn from one set of data, with the same number of classes but different sequences of intervals, or with the same type of sequence but different number of classes, are often anything but alike; this is certainly a weakness of isarithmic mapping. This fact was recognized by Ripley as early as 1898 for the choropleth map and was demonstrated later by both Wright and Mackay for isarithmic maps, and by Jenks for choropleth maps.[18] Thus, when the cartographer determines the number of classes and the type of interval, he selects only one of a large number of possible ways of characterizing a statistical distribution; his reader has no opportunity to see either the original data or alternative ways of representing them. With an appropriate choice of interval the cartographer can effectively characterize the distribution; with a poor choice he can seriously misrepresent it.

Error Characteristics of the Isopleth Map

The isopleth method involves many possibilities for error that are inherent in the mapping procedures, as well as for those that derive from the kinds of data upon which the maps are based. If a cartographer possessed complete knowledge of the details of the distribution which he maps by isopleths, he would be able to check his map against it. For example, he could measure the map errors (if there were any) by obtaining differences between the z values of his map and those of the original distribution. In reality, of course, no one can catalogue all probable errors and sources of error which are involved in the process of isopleth mapping. The procedures which are believed to introduce significant errors are here designated as error characteristics of the method.

Studies of the error characteristics of isopleth mapping have appeared infrequently, considering the widespread use of this technique. Since the 1950s, there has been a growing interest in the inherent errors of isopleth mapping, however, and this has resulted in an expanding literature. A few of these studies attempt to evaluate quantitatively certain sources of error arising from the data and/or the mapping procedures, and then propose some compensatory measures to reduce the error. Most of the recent studies, however, are only preliminary in that they merely point out in broad terms some probable error sources. Further studies are needed if the error characteristics are ever to be measured and controlled, assuming that this is possible. In the remainder of this chapter we shall discuss four of the major sources of error in isopleth mapping: the quality of the data, the system of control, interpolation, and the nature of the unit areas.[19]

The Quality of the Data. The first step in any quantitative mapping is, of course, the collection of data, and the quality of the data directly affects the accuracy of a map. Blumenstock has pointed out that three types of errors are involved in the z values of the data used for isarithmic mapping, namely, observational error, bias error, and sampling error.[20] Only the sampling error, which is closely related to the present research, will be discussed here.

There are usually at least two kinds of sampling errors in the z values used for isarithmic mapping; one concerns time and the other space. Many maps are constructed for the purpose of characterizing a general condition by utilizing quantitative data obtained at particular instants or periods, for example, population as recorded on April 1, 1967, or average farm income during the period 1960–65. With the notable exception of the land surface, all other quantitative distributions shown on isarithmic maps are sample arrays with respect to space. When the land surface is mapped photogrammetrically, the data of the entire population of elevations are theoretically available. On the other hand, a map of mean July temperatures is prepared from the records obtained at a limited number of places. Its isarithmic (but not isoplethic) representation is obviously that of a sample surface, since the total population is the infinite number of point-temperatures in the area concerned.

In isopleth mapping the z values cannot occur at points but are enumerated within defined areas. The usual practice in such mapping is to utilize data collected with reference to either administrative units or the units of an arbitrary pattern designed by the cartographer. In any case the particular system of area subdivision is only one of many theoretically possible systems that might have been adopted. For example, had the historical development of administrative boundaries in an area been very different from what it has actually been, then the data would be collected upon a different set of administrative units.

For a given area and a given distribution, the census data are enumerated according to defined subareas, but if the boundaries of the subareas were altered and new data collected upon the new system, a different set of numbers would result. In other words, for a given area and a given distribution, one system of areal subdivision will provide one set of data (that is, one set of x, y, and z values), and another system, another set of data.[21] It is in this sense that any isopleth surface is a sampling surface.[22] Moreover, since what we know of the quantitative characteristics of the distribution can be obtained only through the enumeration of the data by way of the subdivisions, a certain amount of "sampling error" is introduced due to the variations of the unit areas.

Blumenstock estimated the magnitudes of three kinds of error in the z values of temperature data, and he then proceeded to adjust the original data on the basis of these estimates.[23] The final map, drawn from the adjusted data and considered to be a "proper picture of the distribution," is indeed much generalized, with relatively smooth lines and no isolated enclosed isotherms, in marked contrast to the map prepared from the unadjusted data. Although Blumenstock's study focused on isometric lines, his valuable conclusions certainly can be applied to isopleths as well. Many of the angular lines and small islands on isopleth maps are probably due to errors in the z values and to the various assumptions in isopleth mapping. Therefore, they should be smoothed out rather than left on the map to give an impression of a higher accuracy than the data warrant.

The System of Control. After the data collection, the next step in manual isopleth mapping is the establishment of a system of control, that is, the set of control points in the horizontal plane. The data used for isometric lines, such as spot elevations or temperature records, can readily be assigned to definite point locations, these being the places where the data were obtained. On the other hand, the locations of the control points used for isopleth mapping are not so simply defined, and error is likely to be involved in the process of positioning them. For example, the calculated density of population of an enumeration district is of course an average value for the entire area. Since the control point value must be located somewhere to provide the "spot heights" for interpolating the positions of the isopleths, the question arises: should it be considered to exist at the "center of area" or somewhere else? In order to make that decision the cartographer must assume either evenness or unevenness concerning the distribution of population within the district.

Although absolute uniformity of a geographical distribution probably rarely exists, the cartographer may be justified in assuming uniformity within the unit areas if the larger distribution is fairly smooth, or if the accuracy of the data and/or the scale of the map do not warrant a further breakdown of the data. If the cartographer does assume an even distribution of the population within the unit, then the positioning of the control point is simplified: the average density of population of a district is placed at the center of the district, which is the areal center as well as the assumed center of gravity.[24] In some cases, however, the cartographer may feel that the variation within the unit is such that he cannot assume an even distribution, and it would then be inappropriate to locate the control point at the areal center. In an uneven distribution, the point of the mean, that is, the center of gravity, should be used as the control point.[25] It must be noted, however, that even when control points are carefully located at centers of gravity, these locations are correct only within the level of generalization of the data. Mackay, while recommending the use of the center of gravity, also points out two other centroids as alternative control point locations: the intersection of a pair of perpendicular bisectors and the point of minimum aggregate travel (the median point) of the distribution.[26]

To summarize: In isometric line mapping the distribution of the control values is prescribed and therefore the process of locating the control points introduces no error to the map, but in isopleth mapping the situation is quite different. Isopleth data, being aggregated values referring to enumeration districts, cannot occur at points; nevertheless control points are required in the mapping process, and some approximation is necessarily introduced no matter where they are placed.

Interpolation. Assuming that the cartographer has chosen an interval system and has distributed the values according to his system of control points, he must next determine the positions of the isopleths with reference to the locations and the values of the control points. This involves interpolation, a process employed in all isarithmic mapping which uses control point values and which can be applied only if the mapped surface is assumed as being continuous. Interpolation can be accomplished graphically or mathematically, by hand or by machine.[27] If it is done by machine, one must select an interpolation model for fitting to the array of sample points by the computer. There is no limit to the number of algebraic interpolation models possible and there are many problems involved in their selection and application.[28] In general, computer mapping of isarithms has so far been applied mostly to isometric mapping, probably because of the problems connected with control point location in isopleth mapping.

The elements associated with interpolation in isopleth mapping are most clearly detailed by considering the problems involved in locating the isopleths by hand — the way in which nearly all such mapping has been done to the present. From the "vertical" differences of the control values and the horizontal distances separating the control points, one assumes a gradient along the continuous surface. On the basis of this gradient, the intermediate z values along the surface can be estimated, and the isopleths can in turn be located.

What is the exact gradient and surface variation between a pair of control points? There is no way of knowing; but in order to proceed with the process of interpolation, it is necessary to make some assumption regarding the surface variation between pairs of control points. Thus one may assume a constant rate of change along the line between the two points — that is, a linear gradient — or one may assume some nonlinear rate of change, in which case the gradients would be concave or convex curves. There is, however, no justification for assuming anything more complex than a simple curve of second degree and first class between any pair of control points, since only the values of the end points of the line are known.

In most cases the assumption of a linear gradient is adopted, and the process of scaling off the intermediate values between the end points is called linear interpolation. When the cartographer employs linear interpolation, the statistical surface is actually transformed into a surface consisting of a series of planes (Figure 2).

There is, however, some uncertainty associated with the methodology of isopleth interpolation. Both Uhorczak and Mackay have pointed out the classic problem of the "dead corner."[29] In some instances enumeration districts are so arranged that four adjacent units have a common corner, a condition especially likely to occur when the units are rectangular. If both the values of one diagonally located pair of control points are higher than the values of the other pair, they present a problem of interpolation, since isopleths could be drawn in either of the two ways shown in Figure 3. The cartographer

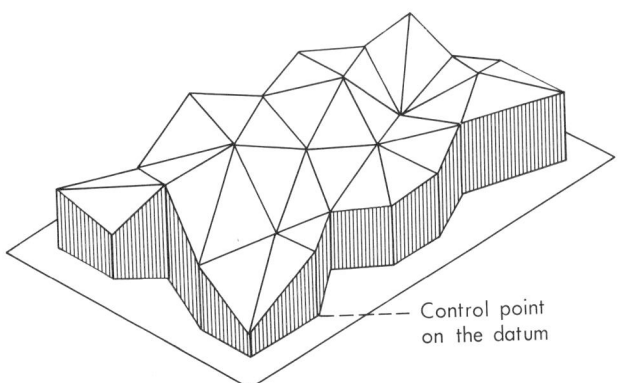

Figure 2. Perspective Representation of the Consequences of the Assumption of Linear Interpolation. The lower flat surface is the datum for the statistical surface above it and is usually the map plane. The height of each vertical above a control point (one per unit area) is proportional to the magnitude of the control point value. (After Schmid and MacCannell.)

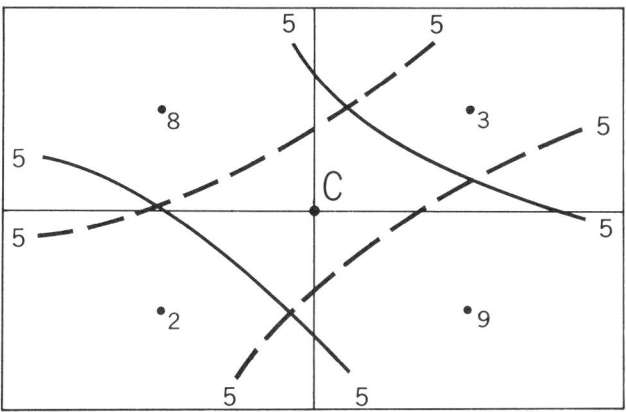

Figure 3. Alternative Choice of Interpolation. The isopleths shown by dashed lines would be chosen if point C is assumed to have a value of 2.5 (between control point values of 2 and 3); the solid lines would be chosen if point C is assumed to be 8.5.

proceeding by the hand method is faced therefore with an alternative choice of interpolation. The problem can be avoided, however, by transforming the data to a pattern of unit areas that does not allow this to happen, such as staggered rectangles, rhombuses, or hexagons. Control points in the centers of such an array of shapes are triangularly arranged and thus present no problem of alternative choice.

Linear interpolation assumes the simplest possible gradient, a straight line, between a pair of control points. Logically, if one knows nothing but the end values this assumption is the most appropriate because it is the simplest. Nevertheless, most geographical distributions are known to have variable rates of change per horizontal unit, that is, to have curvilinear gradients, and a certain amount of inherent error is involved in the use of linear interpolation for isarithmic mapping. The amount of such error increases directly with the curvature of the actual gradient between any two points. Porter has illustrated the displacements of isopleths that are introduced by linear interpolation along a concave slope.[30] If the curvature is modest, or if the control point density is high, the error so introduced is relatively not large.[31] Moreover, most distributions probably have compound curvatures; that is, the surface is likely to be characterized by numerous concave and convex slopes with inflection points between. In such cases, the errors introduced overall by linear interpolation would probably be randomly distributed rather than persistent or cumulative, since the amounts of the errors would vary with the curvatures, and the directions of the displacements of the isopleths would depend on the types of gradients.

The type of error involved in linear interpolation is inherent in any system of interpolation unless the surface has a constant gradient everywhere. Instead of a linear assumption, for example, one may assume the gradient between control points to be a defined curve of some kind, such as the modal type slope of distribution; nevertheless, some displacement of the isarithms must be expected because not everywhere will the distribution be fitted to this curvature. Therefore the process of interpolation must inevitably introduce some random errors in isarithmic mapping.

Thus far the discussion has been directed to only one aspect of interpolation, namely, the determination of the z values between control points in order to locate isarithms. Actually there is another phase of interpolation — the estimation of the z value at any given point of an isarithmic map after the isarithms have been positioned. This, of course, is not an error characteristic of the mapping process; rather it is related to the "total" commensurability of the isarithmic map.

In view of the probable error involved in the method of linear interpolation, many researchers are inclined to modify the method somewhat in order, hopefully, to reduce its error effect. For example, some cartographers urge that the map maker should not concern himself only with one pair of control points at one time, but should also view the characteristics of the surface as a whole as deduced from all the plotted control point values.[32] Thus if an array of values, spaced approximately equally, are read as 10, 20, 40, 80, and 160, the slope is obviously not constant; rather it steepens in the direction of the highest value. Therefore the linear scale of interpolation should be modified accordingly. In any case the application of a mechanical method, such as linear interpolation, ought to be modified if, by so doing, a higher degree of accuracy can be attained.

For mapping population density in the United States by county data, Alexander and Zahorchak modified the linear interpolation method with Wallis's spot height concept.[33] Uhorczak has discussed several different types of nonlinear interpolation scales, namely, geometric progression, a natural logarithmic, and one applying the logarithm of the sine function.[34]

The method of linear interpolation has been widely

adopted in the past in part, no doubt, because it is relatively easy to use. The situation is quite different, however, when a computer is available and the amount of computation is not a major concern. In linear interpolation the cartographer considers two control points at any one time and assumes the simplest gradient between the points. With the use of a computer he can consider three, four, or all the control points at once and can assume a continuous and undulating surface lying between all the points. The latter approach, at least, has a much sounder theoretical basis, for the mapped surface is assumed as being continuous and undulating from the very beginning of the mapping procedure, whereas linear interpolation involves an assumption that the surface is described by a series of planes joined at the control points and along straight lines between them (Figure 2).[35]

The Nature of the Unit Areas. It was noted earlier that two types of data are employed in isarithmic mapping; one type can be assumed to exist at points and the other cannot. The latter is enumerated upon areal subdivisions. They are called unit areas, and their geographical arrangement is designated as the unit area pattern.

The data from which isopleth maps are made are normally collected from either an arbitrary unit area pattern designed by the researcher or a pattern of civil administrative or enumeration subdivisions. The first type usually involves unit areas with some regular geometric arrangement of boundaries like hexagons, rhombuses, or rectangles. The unit areas of such a pattern are designed to be systematically arrayed, that is, uniform in both shape and size. The unit areas of the second type, on the contrary, usually have variable shapes and sizes and thus an irregular pattern.

If a given volume distribution is subdivided into unit areas for data enumeration, and if the unit areas are of the same size and of some optimum shape, then the use of areal data is an aspect of generalization. The number of unit areas into which the given distribution is divided is largely a question of the level of generalization. In practice, however, as we have seen, most enumerated data are based on variable unit areas; accordingly, the use of areal data involves not only concern with level of generalization but also concern for the error characteristics introduced by the variable unit areas. The variations in size and shape of the unit areas and the irregularity of the unit area pattern are sources of "sampling error," as was pointed out previously, and these variations affect the level of accuracy of the isopleth map.

When data are collected upon unit areas of variable sizes, each observation has its own level of generalization and, in turn, its own probable error of the z value. The map derived from these data necessarily inherits these variations, and therefore there is a high probability that some parts of the map will have a lesser degree of generalization and a closer and more detailed representation of the configuration of the volume distribution than will other parts of the map. Of course, these "false" differences, which are created by the system of area subdivision and data collection, are simply "included" in the map, and it is very difficult, if not impossible, to separate them from the "true" variations of the distribution itself. These differences in level of generalization probably result in lowering the overall accuracy of the map as a whole. The difficulties introduced by variations in the unit areas become even more critical when the reader's attention is not directed, as it rarely is, either to the variations in the sizes of the unit areas or to the uneven distribution of the control points. Consequently, the reader is likely to be unaware of the problem and readily assumes a uniform degree of generalization throughout the map. One way to inform him about these variations, as suggested by Wright, is to present him with the distribution of the control points.[36] This is often done in climatic studies, more or less automatically, by showing the locations of the various weather stations.

The unit areas in a pattern are not only rarely of equal size, they are also seldom regular in shape. A pattern of unit areas of variable shapes and sizes must have an irregular arrangement, and the network of the control points will correspondingly be irregular. Irregularity has several effects, as will be illustrated later in this study. The cartographer can obviate some of these difficulties by designing his own unit area pattern if this is possible in the particular circumstances of his study. Ideally, he can choose a geometric arrangement which provides a triangular network of control points in order to eliminate any alternative choices in his interpolation.[37] Furthermore, haphazardly arranged enumeration data may be transformed before mapping to a desirable regular pattern, such as hexagonal.[38]

The Experimental Design

A major objective of the present research is the investigation of error characteristics that are introduced when the mapped data are enumerated upon unsystematic unit areas (e.g., counties or parishes) of varying shape and size. Such an investigation requires a rigorous study of several variables related to the fidelity with which an isopleth map portrays a geographical distribution. These are: the surface characteristics of geographical volumes, that is, three-dimensional "statistical surfaces"; the variation in the sizes of the unit areas; and the variation in the shapes of the unit areas. Concurrently a fourth element is included, namely, the optimum size of hexagons that may be used in the transformation of data from an unsystematic array of irregularly sized unit areas to a hexagonal pattern.

The term "accuracy" (of a map) has no standard meaning, though some criteria have been developed in the field of topographic mapping, where it is a common practice to check the accuracy of the contours on a map by comparison with the actual land surface.[39] Unfortunately little work has been done on the problem of measuring the accuracy of isopleth maps. The set of isopleths on a map obviously characterizes a three-dimensional surface, but that symbolized surface must remain as a hypothesis to the cartographer because he has derived it from a particular array of z values based upon arbitrary unit areas and, accordingly, he does not know exactly the surface which he is attempting to map with isopleths. As a consequence, he has no way of determining precisely the errors at the various points of his map.

In order to study and evaluate the basic elements in isopleth mapping which affect the accuracy of the map, one must first establish a "total" distribution of z, comparable to the land surface. To this end the authors proceed systematically as follows: (1) a number of original surfaces are selected; (2) a variety of unit area patterns are selected; (3) unit area data for these patterns are derived from the values of z on the original surfaces; (4) the unit area data are converted to hexagonal data; (5) isopleth maps are constructed from the hexagonal data; and (6) these isopleth maps are then compared with the original surfaces. Each of these major steps in the experimental program will be discussed separately in Chapter 2.

2

The Preparation of the Experimental Data

AS IMPLIED at the conclusion of Chapter 1, the accuracy of a map must be arbitrarily defined in some way. There is no standard procedure that is universally followed in such evaluations nor is there any generally accepted definition of the term. This is true even in the field of topographic mapping, where contour location is now largely the result of machine-like photogrammetric methods of high rigor. A rigorous process is emphatically not characteristic of isopleth mapping, where there are some variables over which the cartographer has little or no control.

For the purposes of this study, the term accuracy is defined specifically as the fidelity with which an isopleth map represents the z values of a given volume quantity. It is measured by observing the difference between (1) the values of z of a known continuous original distribution at a number of testing points, and (2) the values of z as interpolated at the same points on test isopleth maps prepared from the "same" distribution of z when the z values have been aggregated by unit areas in a variety of ways. In order to simplify the writing, the group of z values from the original distribution will be called the "original values" and those from the isopleth maps, the "map values." It is likely that, in addition to the given statistical information to be mapped, a cartographer will have some familiarity with a variety of the characteristics of the distribution which he is mapping. This information may be very valuable since it can serve as a guide for some of the decisions the cartographer usually needs to make throughout the whole mapping process. Such knowledge will ultimately affect the quality of the map, but it is neither measurable nor a cartographic variable.

Selection of the Original Distributions

Since one must have a completely known reference surface in order to evaluate the effects of various mapping methods upon map accuracy, four such surfaces, that is, distributions of z, were chosen. It would have been most desirable, of course, to select distributions whose characteristics permitted the generalization of conclusions to a large group of the kinds of geographical quantities commonly mapped by isopleths. It is as yet impossible, however, to choose "typical" geographical quantities; although there are numerous types of distributions, many of which have been mapped by isopleths, a systematic classification or grouping of them according to surface characteristics has not been established. Conceivably this is possible, and one such system of classification might even contain several polynomials, each describing a certain type of distribution. Such research would certainly be useful, but it is obviously too ambitious to have been part of the procedure for the present study.

The main criterion in the choice of the four original surfaces is that each surface should be clearly differentiated from the others with respect to its distributional pattern and its surface complexity. These characteristics are not numerically defined; on the other hand, the surfaces are obviously distinguishable when described by isarithms with an interval of one on a datum of zero z value (Figures 4 to 7). The complexity clearly increases from Surface I to Surface IV, and one of the hypotheses to be tested is that, if other variables are held constant, the simpler the z surface the more accurate will be the resultant isopleth map.

Surface I is a series of concentric arcs (Figure 4). Its equally spaced isarithms portray the constant gradient of a portion of a cone, and it has a "local relief" of 38, this being the difference between the highest and lowest points within the map. This surface, the least complex of the four, serves as a kind of control in relation to the other three. Surface I is, of course, only a geometric form, and no comparable surfaces are to be found among geographical quantities which might be mapped by isopleths.

Surface II is also relatively smooth, though it shows irregularities of moderate complexity and a local relief of 39 (Figure 5). The highest point is in the upper right corner and the surface slopes away in all directions from this point. Generally speaking, the gradient decreases

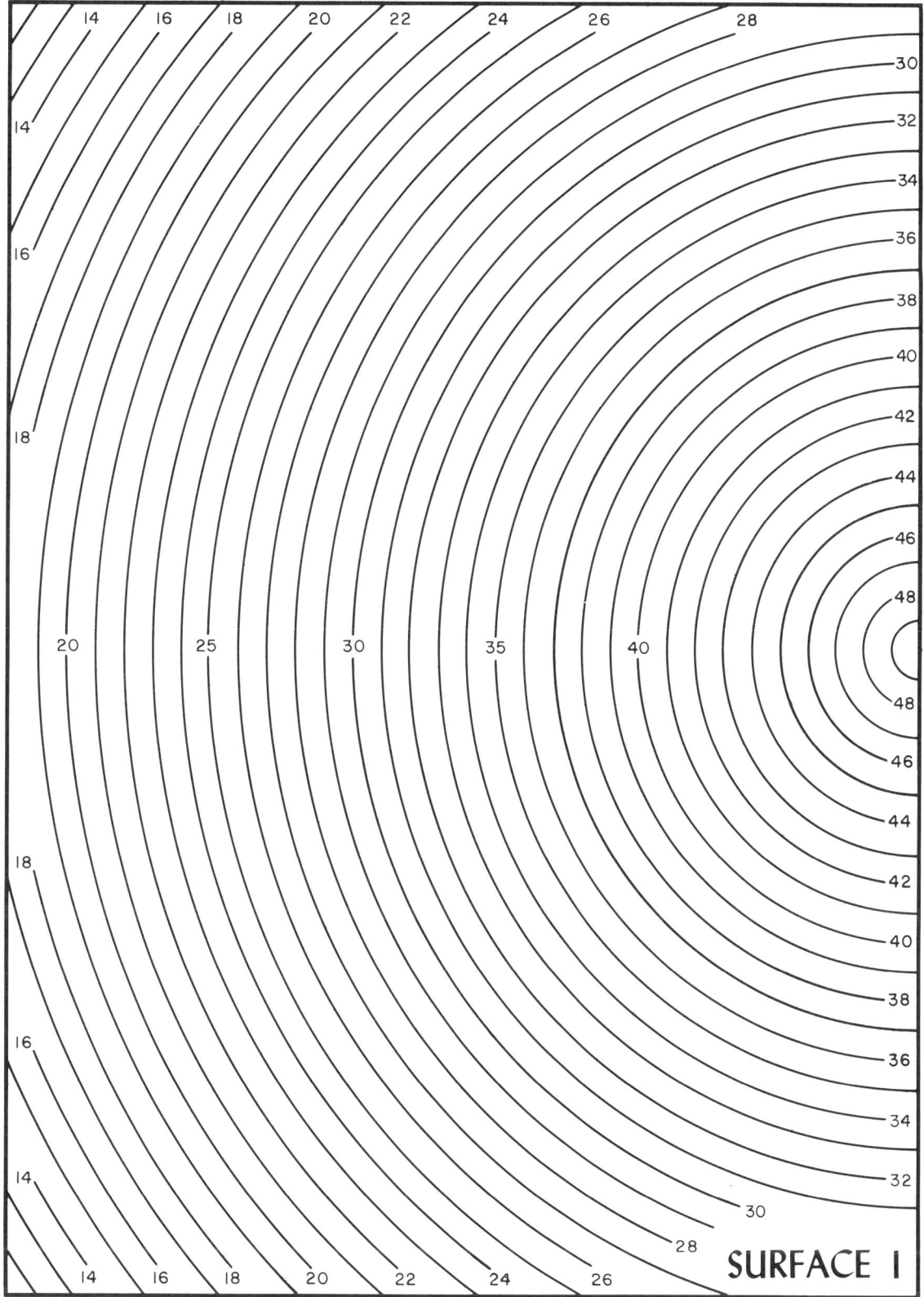

Figure 4. Original Surface I

Figure 5. Original Surface II

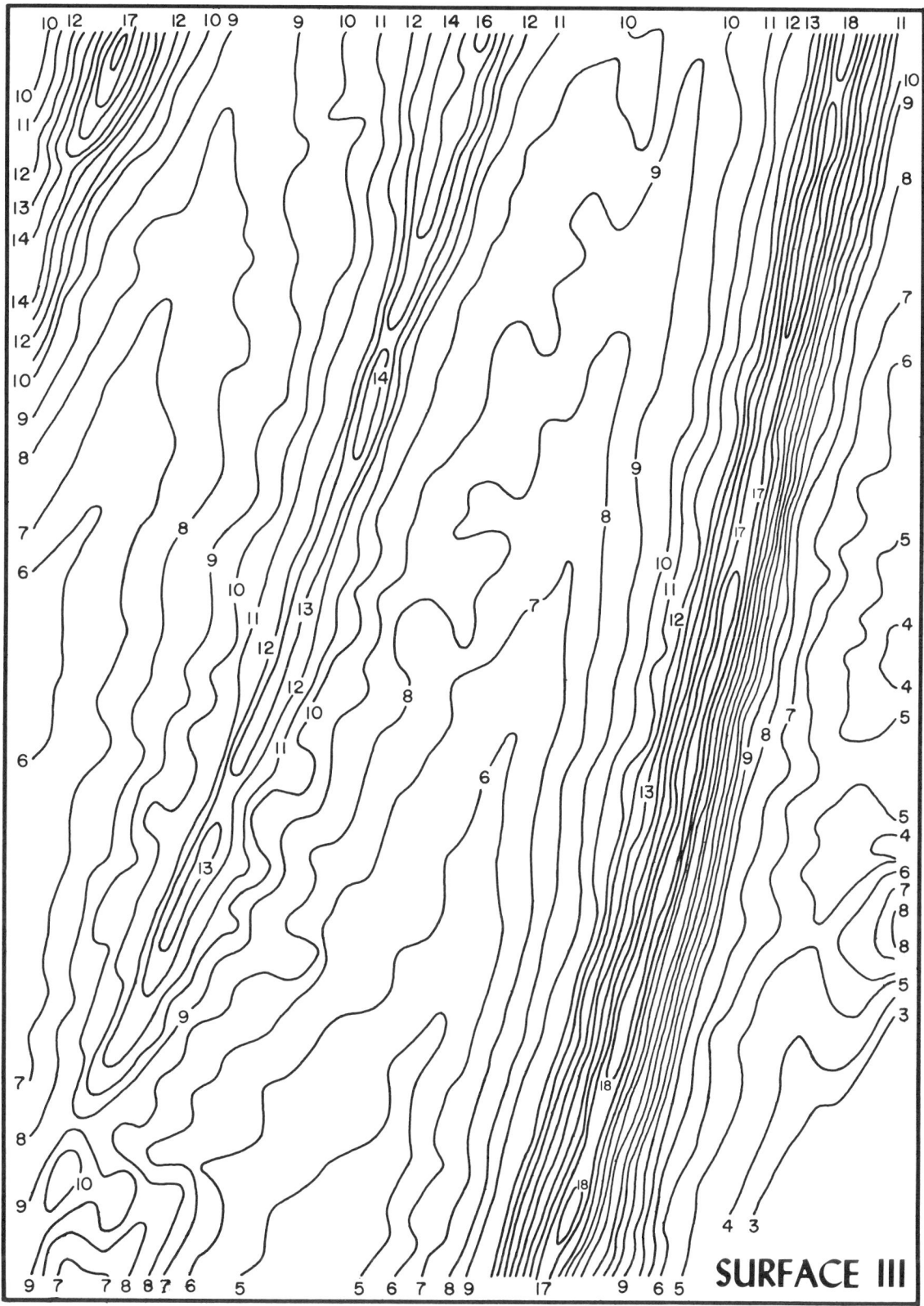

Figure 6. Original Surface III

Figure 7. Original Surface IV

gradually from the upper right to the lower left. Surface II symbolizes a distribution similar to one-fourth of an irregular cone, and geographical volume distributions comparable to a total surface of this kind are readily found, for example, the distribution of newspaper circulation in and around an urban center.

Surface III is more complex, with marked linearity, although the local relief is only 16 (Figure 6). Its choice is based on the assumption that such a total distribution of z would be fairly difficult to portray by isopleths; one would need a large number of sample points in order to preserve the directional characteristics of the distribution. Many geographical quantities have distinctive linear concentrations, although the concentrations may not be so regularly localized. Surface III resembles the distribution of soil or land-use characteristics in an area with a marked linear "grain," for example, beach ridges or ridge and valley terrain.

Surface IV has a large amount of spatial variation, and thus a high complexity; however, it has no distinctive distributional trends (Figure 7). Its local relief of 20 is relatively small. One reason for its selection is to assist in the understanding of the general relation between the complexity of a surface and the accuracy of the resultant isopleth map. Such complex and irregular distributions are rather common: an example might be the distribution of the proportion of population using a certain language in a region of mixed ethnic origins.

Surfaces II, III, and IV are actually somewhat simplified contour representations of the land surface taken from topographic maps. For our purposes, however, the four surfaces are assumed to be abstract; thus the isarithms which depict them are only numbered and do not represent any absolute units. The isarithms portraying the surfaces are constructed within a uniform dimension of 8 inches by 11 inches, and they are assumed to be of like scales.* The scale is not defined because the study is entirely analogical; here again absolute units are not pertinent.

Selection of Unit Area Patterns

In addition to the four surfaces which serve as the "given" real distributions (for example, density of population), four patterns of unit areas simulating administrative subdivisions are chosen in order to study the

---
* The dimensions as stated here refer to the actual size used in the experiment. The illustrations in this volume have been somewhat reduced.

effects that variations in the patterns of unit areas may have on map accuracy. Each pattern has distinctive characteristics in both the shapes and the sizes of the subdivisions. The shape characteristics are selected by eye, that is, they are visually distinguishable from one another; but the size characteristics can be quantitatively described.

The four patterns are drawn the same overall size as the four surfaces, 8 inches by 11 inches. Pattern I is composed of areas of uniform shape and size (Figure 8); Pattern II is composed of areas with fairly comparable shapes but of varying size and orientation (Figure 9); Pattern III is composed of areas with varying shapes but almost uniform sizes (Figure 10); and lastly Pattern IV is composed of areas that vary in both shape and size (Figure 11).

Table 1. Average Size and Variances of Unit Areas

Pattern	Number of Unit Areas	Average Size (square inches)	Variance (square inches)
I	30	2.93	0.00
II	30	2.93	1.69
III	30	2.93	0.53
IV	31	2.84	2.11

Table 1 shows the average unit area sizes and their variances. The subdivisions of Pattern III vary somewhat in size but the variance is small in comparison with Patterns II and IV. The results of F tests between the variances of Pattern III and Pattern II, and Pattern III and Pattern IV, showed the differences to be significant at the 1 percent level. In the later analysis Patterns I and III are grouped as having equal-sized unit areas, in contrast to Patterns II and IV.

Pattern I represents the most regular form of statistical subdivision, while the other three patterns are characteristic of the kinds and arrangements of enumeration districts where most census activities in the United States have been carried out. As a matter of fact, Patterns II, III, and IV are modified from maps of county boundaries at various locations in the United States.

Derivation of Unit Area Data

After the selections of the four surfaces and the four unit area patterns, the next step in the research is to obtain the experimental data, namely, the averages of the z values occurring in each unit area of a pattern when a

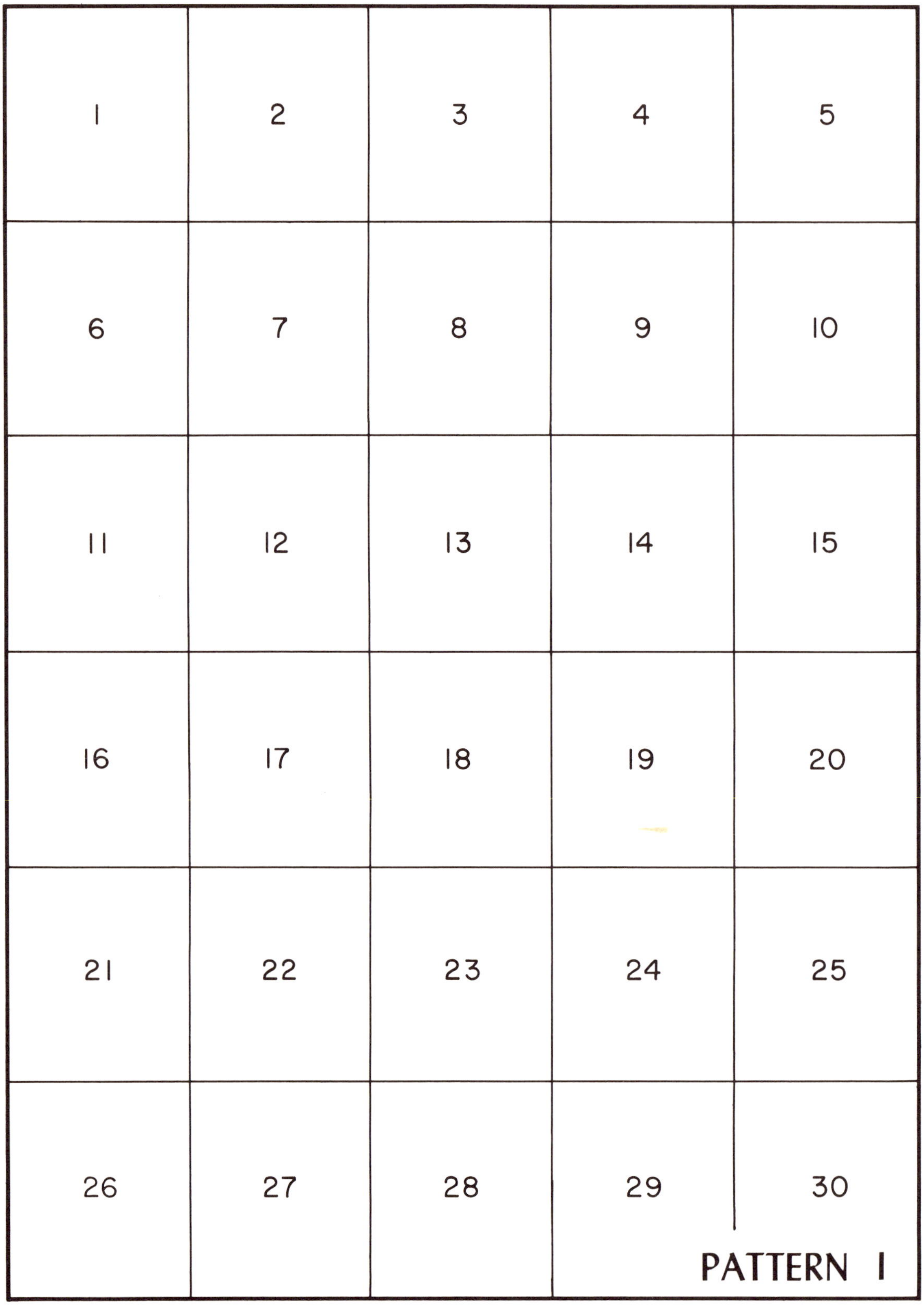

1	2	3	4	5
6	7	8	9	10
11	12	13	14	15
16	17	18	19	20
21	22	23	24	25
26	27	28	29	30

PATTERN I

Figure 8. Unit Area Pattern I

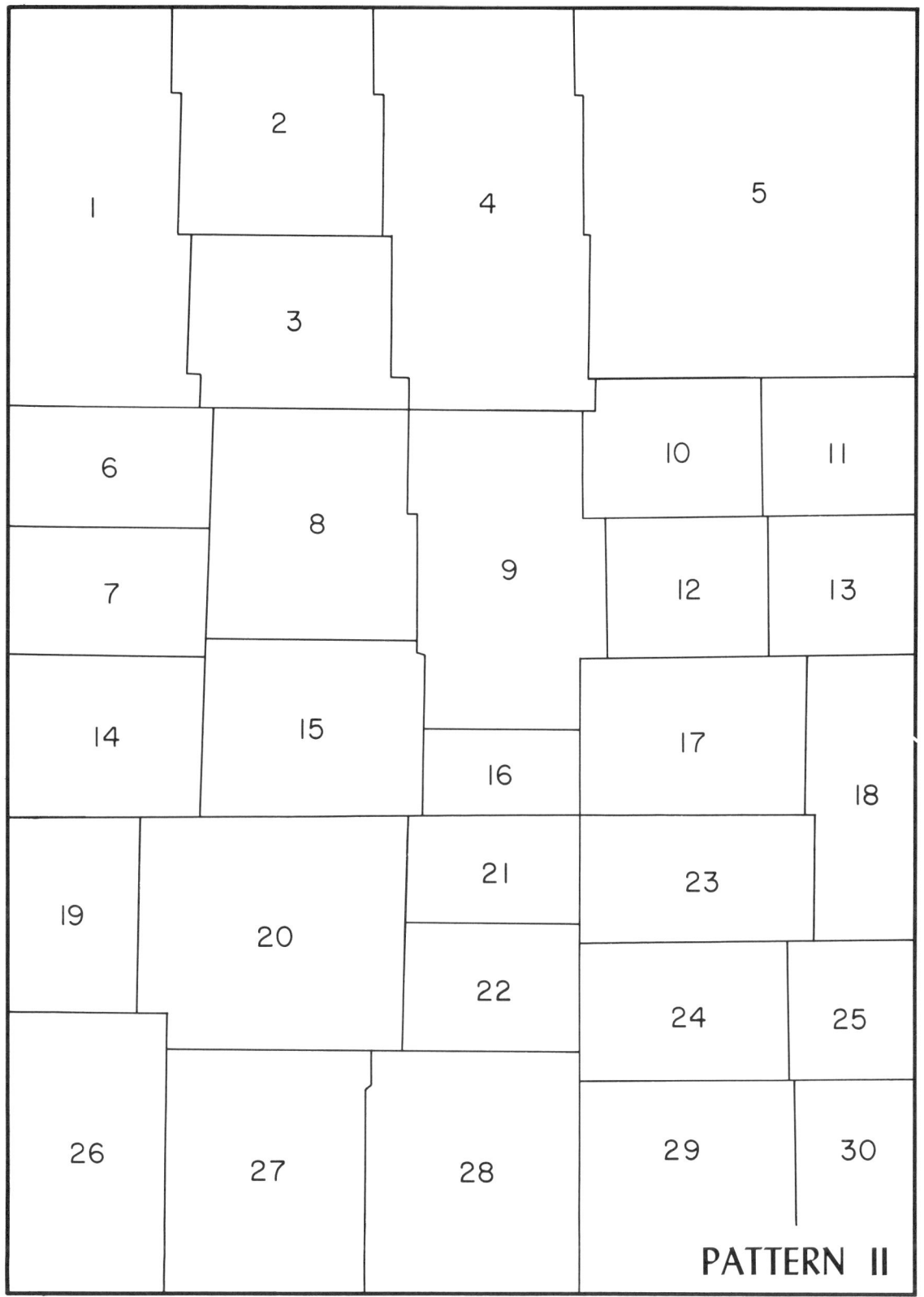

Figure 9. Unit Area Pattern II

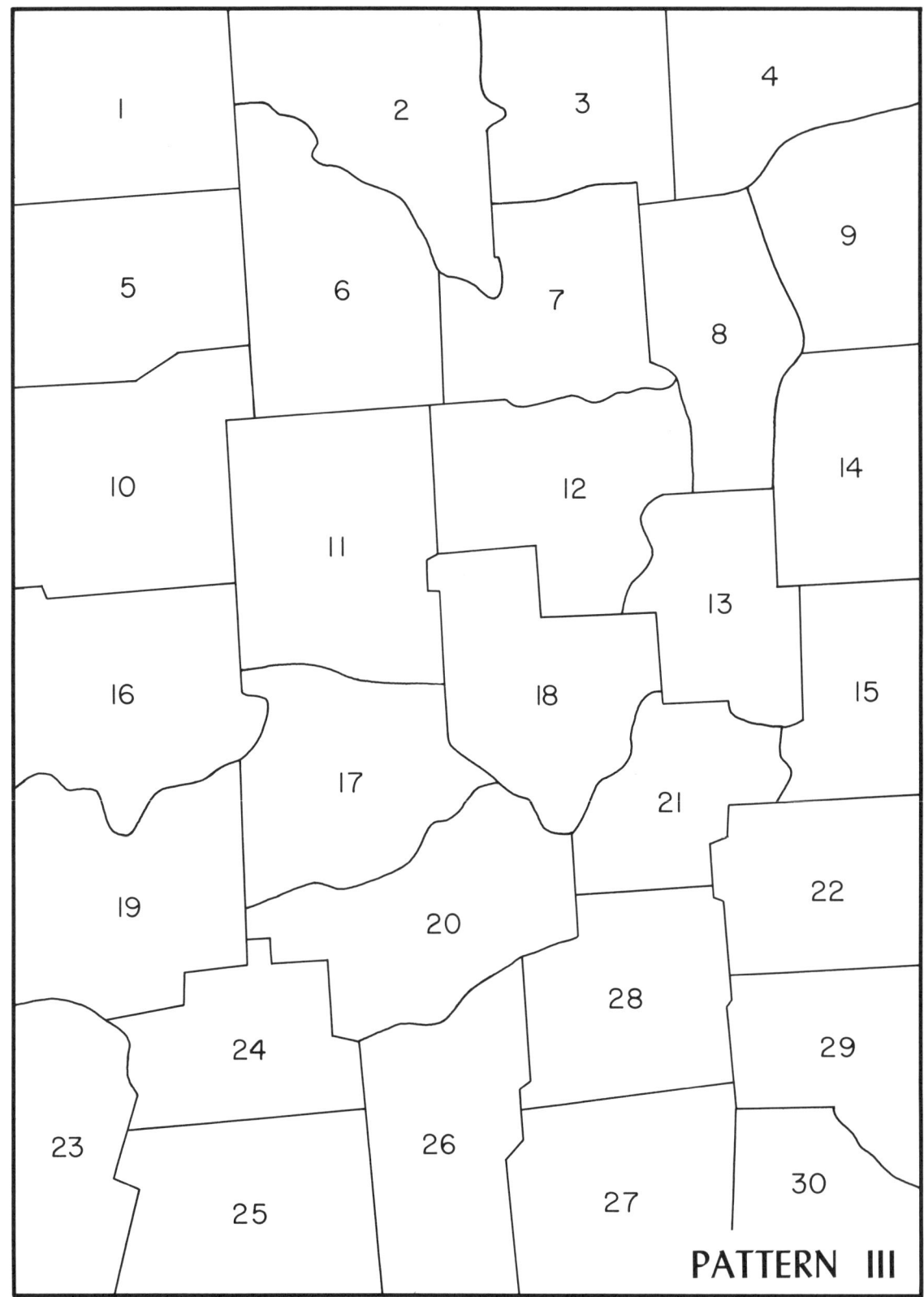

Figure 10. Unit Area Pattern III

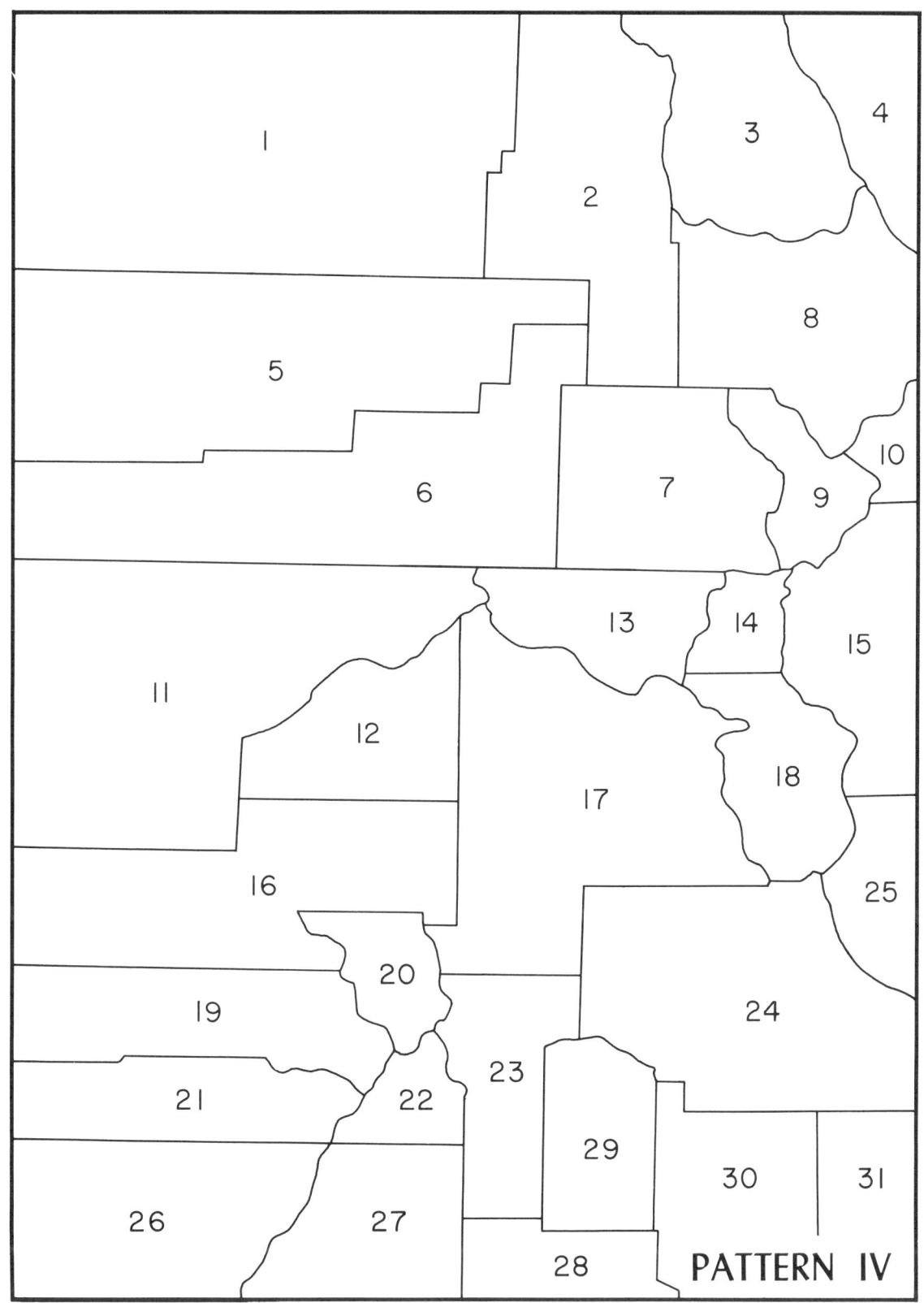

Figure 11. Unit Area Pattern IV

pattern comprises the set of enumeration districts for the area of a distribution shown by a surface. Each pattern is in turn superimposed upon each surface to form the sixteen possible surface-pattern combinations; in this way each given distribution is enumerated four different ways. A set of discrete unit area data consisting of mean z values is then derived for each of the sixteen combinations.

A geographical mean of the z values for each unit area (enumeration district) within a pattern is obtained by measurement and is calculated as follows (see Figure 12):

Unit area value $= U = \dfrac{\Sigma a_i m_i}{A}$

where $A =$ the total area of the unit area
$a_i =$ the area of the ith portion of the unit area between a pair of isarithms,
$\therefore \Sigma a_i = A$
$i = 1, 2, \ldots, (n + 1)$
$n =$ number of isarithms within the unit area
$m_i =$ mid-value of the pair of isarithms bounding the ith portion of the unit area.

All measurements are made with a polar planimeter to a hundredth of a square inch, checked, adjusted to conform to known totals, entered on individual worksheets, and processed with a desk calculator. Because of the checking procedure, any significant error is highly unlikely, and the tiny errors that result from the inability of the planimeter to record to less than a hundredth of a square inch will be randomly distributed.

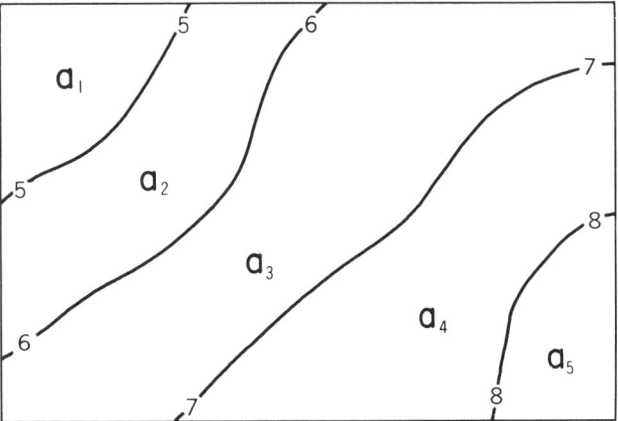

Figure 12. Derivation of a Geographical Mean for a Unit Area from the Isarithms of a Surface. (See text for further explanation.)

A value so derived is actually an approximation of the "true" mean, since a mid-value between adjacent isarithms is used to represent the infinite number of point values which occur within the isarithmic zone. As is evident from a study of Figure 12, for example, if one were to assume the simplest surface gradient, employment of the mid-value would probably result in an underestimation of the value of a_1 and an overestimation of the value of a_5. In most cases, such errors will normally cancel one another, since they are likely to be randomly distributed. In a few cases, however, the area of a_1 is persistently larger than that of a_5; thus a larger area is being underestimated than overestimated. In any event the errors so introduced can have little effect on the results of the total experiment.

Transformation to a Hexagonal Pattern

The entire isopleth mapping procedure, from the collecting of the data to the final drawing of the map, involves many operations, each of which may affect the fidelity with which the map portrays the real characteristics of the distribution. For example, any chosen interpolation procedure will always affect the accuracy of the map because any assumptions, if they are systematic, are only approximations of reality.

This study does not attempt to evaluate all the elements involved in the accuracy of isopleth mapping; instead it focuses on the effects which result primarily from unit area variations and secondarily from different surface characteristics. Because it is a sampling study, elements other than the ones being studied should ideally be maintained as constants throughout the entire mapping procedure. If this can be done, one may safely assume that the variation among the samples is due to the variable effects under study and to the sampling error.

It is difficult to design a useful experimental study in cartography in which all elements of variation are completely controlled. In this study several measures are adopted in order to achieve as much control as possible of the elements to be held constant. For example, the derivation of unit area data is carried out with extreme care so that the inevitable errors will be randomly distributed and their magnitude minimized. Similarly, the isopleths of the resultant maps are not located on the basis of the z values of unit areas in the four patterns, but instead the experimental unit area data are transformed to mapping units of uniform size and shape ar-

PREPARATION OF THE EXPERIMENTAL DATA

ranged so that control points will be arranged triangularly, a procedure designed to preclude the occurrence of the problem of alternative choice in isopleth location. A regular hexagonal pattern has been selected for this purpose.

Again, for better control of the sampling procedure, the spatial variation of z within each unit area (or each hexagon) is assumed to be uniform; that is, the control point value which represents the whole unit is located at the center of the unit. The hexagonal pattern is selected because the average distance from the center point of a hexagon to its perimeter is less than in a rhombic, square, or rectangular figure of equal area. Hence the control point for a hexagon is less influenced by values in extreme portions of the area it represents, and therefore, theoretically, it should be a more representative value. The hexagon is the practical alternative to the theoretically best shape to be represented by one point, a circle. Cartographically, the circle is undesirable for mapping since a pattern of circles will necessarily overlap if it is stipulated that they must include the entire mapped area. The allowing of overlap would result in the values for some areas being enumerated two or more times.

A second problem associated with the choice of the transformation unit is its size. In previous studies which employed hexagons for similar purposes, the sizes were more or less arbitrarily selected on the basis of the unique conditions involved.[1] The optimum size, under given mapping conditions, has not been systematically studied, and therefore a set of five different sizes are chosen in order to evaluate the effect of hexagon size on the fidelity of the isopleth map. Since the size ultimately determines the number of hexagons that will occur within a given map area, it also determines the number of control points. By employing a progression of sizes, it is also possible to observe the effect of the variable numbers of control points.

One feels, intuitively, that when transformation is contemplated the optimum hexagon size should be the average size of the given original unit areas. It can be argued that larger hexagons will result in a higher level of generalization and correspondingly the loss of some details of the original unit area data. On the other hand, the use of smaller hexagons may create some spurious detail (cartographic noise). Accordingly, the average size of the unit area was chosen as the mid-size hexagon and four other sizes in simple proportions were selected (Table 2).[2]

Table 2. Sizes of Hexagons

Hexagon	Hexagon Size (square inches)	Ratio to Hexagon C
A	0.73	1/4*
B	1.45	1/2
C	2.90	1
D	5.80	2
E	8.70	3

*Approximately.

Regular hexagon patterns of the selected sizes are drawn on translucent materials, and each of the five hexagonal patterns is superimposed on each of the sixteen combinations of surface and unit area patterns. The system of measurement and calculation employed to make the transformation from the unit area data to the hexagonal data is essentially the same as that used to obtain the geographical means for the unit areas.

The basic unit area data are derived, as described previously, from sixteen combinations of surfaces and unit area patterns. For each unit area pattern, four sets of data are derived, one from each of the four surfaces. Each of the four sets for a pattern describes a unique discrete distribution because the z values of the unit areas are different from one distribution (surface) to another, although the boundary patterns of the enumeration districts in one such set of four are identical. Accordingly, the sixteen sets of data can be considered to be independent statistics. When these data are transformed into a hexagonal pattern, their independence must be retained, and consequently it is necessary that each of the superimpositions of the hexagonal patterns on the unit area patterns be oriented differently. Therefore each of the five sizes of hexagonal subdivisions is superimposed randomly on a unit area pattern four times, once for each combination of pattern and surface.[3] These randomly oriented transformations result in different distributions of control points for mapping. Consequently, the hexagonal data and control points of one transformation are in no way related to any other set and its control points.

Since there are four surfaces, four unit area patterns, and five sizes of hexagonal subdivisions, there is a total of eighty combinations. Each combination will hereafter be designated by a code consisting of the number of the surface, the number of the unit area pattern, and a letter

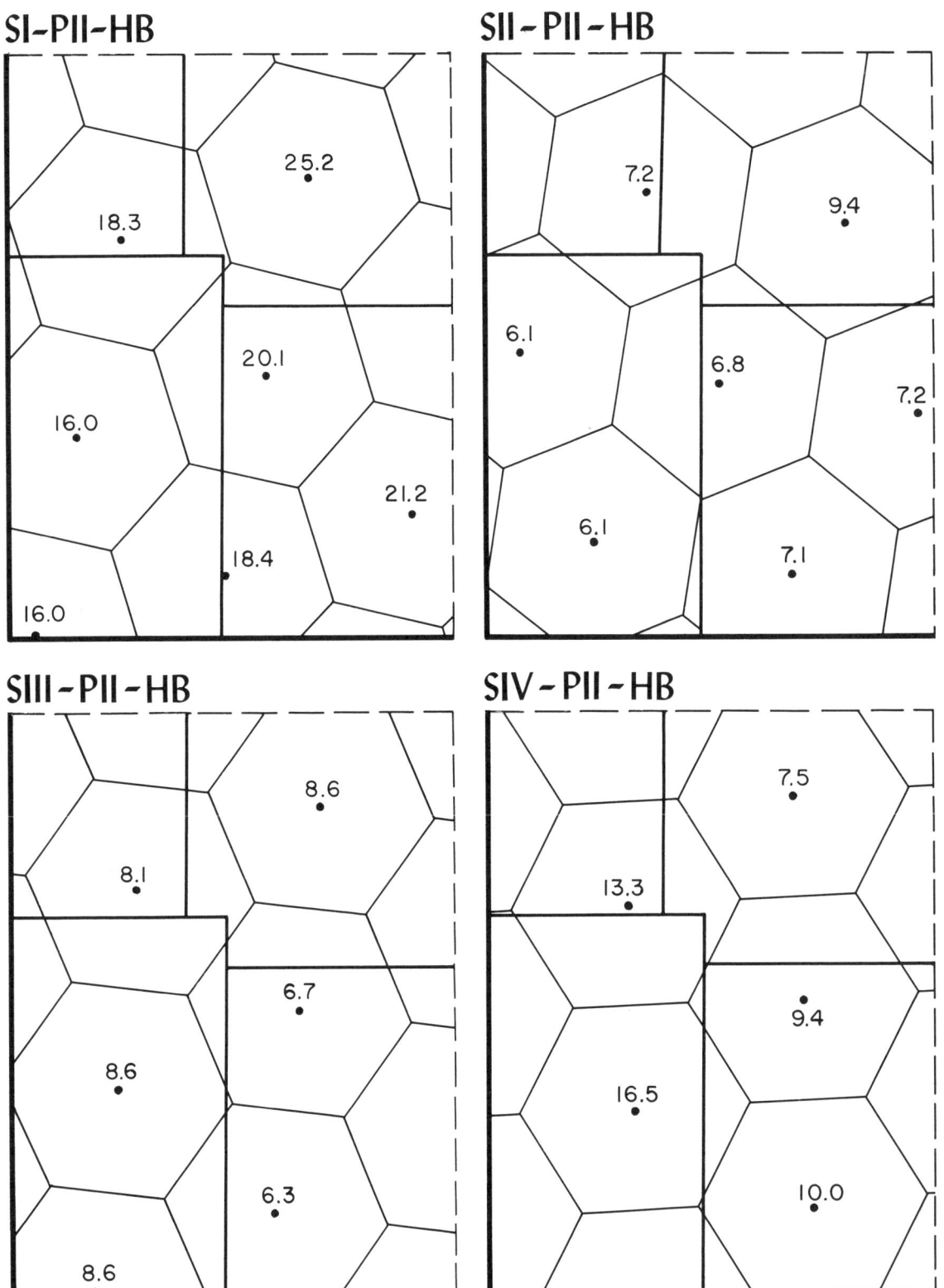

Figure 13. Orientation of Hexagon B on Pattern II for the Four Original Surfaces. From this procedure the data of maps SI-PII-HB, SII-PII-HB, SIII-PII-HB, and SIV-PII-HB are derived. Only the lower right corner is shown.

designating the hexagon size. For example, Figure 13 illustrates the orientation of Hexagon B on (unit area) Pattern II for each of the four surfaces. Accordingly, these four are coded as SI-PII-HB, SII-PII-HB, SIII-PII-HB, and SIV-PII-HB.

A rectangular figure cannot be subdivided neatly by a hexagonal pattern; that is, regardless of how the hexagonal pattern may be superimposed there will always be some hexagons intersected by the outside borders of the entire unit area patterns. In order to be consistent, any divided hexagon whose center falls outside the borders is ignored; all others are treated as complete hexagons in the sense that their centers are considered as control points. Since the superimposition is random, and the acceptance or rejection of a hexagon depends on the position of its center, the number of hexagon centers falling within the borders varies slightly for each transformation. The difference in the number of control points is only one more or less; furthermore, such variations themselves will normally be randomly distributed and will not be likely to affect decisively the character of the isopleth maps.[4]

The foregoing discussion has stressed the purpose and advantages of transforming the data from unit areas to hexagons for experimental purposes. There is, however, a disadvantage in this procedure. In the conversion of mapped data from irregular unit area sizes and shapes to a regular pattern of sizes and shapes (hexagons), the variation of generalization inherent in the unit area data is reduced or "averaged out." Although we are working with "three-dimensional" distributions and the generalizing process is quite different, the effect of the transformation is similar to that obtained by applying a moving average to the irregularities along a curve.[5] The reduction of the variation is likely to be greater in patterns containing large contrasts of unit area sizes and shapes, such as Pattern IV, and less in the more regular unit area patterns, such as Pattern II. Since the purpose of this study is to measure, analogically, the effect of those variations inherent in unit area patterns on the fidelity of the isopleth map, the use of hexagons as an intermediate mapping procedure does introduce a somewhat negative effect.

The Resultant Isopleth Maps

The eighty combinations of surfaces, unit area patterns, and hexagon sizes provide the basic areal data for making the resultant maps. For each of these eighty experimental maps, hexagon values are plotted at their centers, that is, at their control points, and isopleths are drawn. In order to be as consistent and objective as possible, the isopleth values between pairs of adjacent control points are located by strict linear interpolation.

The isopleths based on the data derived from Surfaces I and II have an interval of two, while those from Surfaces III and IV have an interval of one. With the smaller interval, the surface is characterized by twice as many isopleths and thus a larger amount of surface variation can be preserved. It is expedient to use the larger interval for the experimental maps resulting from Surfaces I and II because here the distribution varies gently and regularly and can be efficiently depicted with fewer isopleths. When it is desirable, however, supplementary isopleths (shown with long dashes) of odd numbers are drawn on the resultant maps of Surfaces I and II. They are usually found in the marginal areas of the maps (see Plates 1–16).

Sometimes there are not enough control points along the margins of the maps to indicate precisely where the isopleths should be located; this, of course, happens much more often with the patterns of larger hexagons. In such instances the isopleth is extended to the border of the map with short dashes to suggest the uncertainty of the location. The isopleths are smoothed somewhat to avoid an angular appearance; however, the smoothing in no case alters the basic directions and characteristics of the gradients and can have no significant effect on the subsequent analysis. These isopleth maps, being re-creations of the given distributions (surfaces) obtained by way of subdivision, averaging, and transformation, are designated hereafter as resultant maps in order to distinguish them from the original maps.

3

Testing and Sampling Procedures

AS WE have pointed out earlier, the measurement of the relative ability of an isopleth map to reflect the characteristics of the original statistical surface, after they have been passed through the filters of unit area aggregation and hexagonal transformation, depends upon comparing the z values of the original surfaces with corresponding values on the isopleth surfaces which have been mapped after aggregation and transformation. Not many methods for doing this have heretofore been employed, and most have dealt with the representation of land-surface form by the use of contours (isometric lines) on topographic maps.

Evaluation of Isarithmic Accuracy

The National Map Accuracy Standards for the United States, adopted in 1947, provide a set of specifications upon which the qualities of the nation's large-scale topographic maps are judged.[1] The testing of each map for hypsometric accuracy involves comparisons between: (1) the elevations at a number of points selected along several traverses as interpolated from the map, and (2) the elevations of the corresponding points on the ground determined by field surveys of a high level of precision. The difference between (1) and (2) at any point is designated the error at that testing point. The map is considered to have met the national accuracy standard if 90 percent of the points tested show errors smaller than one-half the contour interval.

European countries use similar "point checking methods," but the map errors are usually not evaluated by reference to the contour interval; instead a "standard deviation" is calculated from the errors by the following formula:

$$d = \left[\frac{\Sigma e^2}{n} \right]^{1/2}$$

where d = the standard deviation
e = the error at a testing point
n = the number of points tested.

The standard of map quality is given by a magnitude of d which in effect serves the same function as the United States standard.[2] As a matter of fact, the United States standard can be easily converted and expressed in terms of a standard deviation.[3] It should be noted, however, that in practice neither the standard deviation nor the method used in the United States alone specifies the vertical accuracy. Since a horizontal displacement of a point on a map will normally cause a change in the magnitude of the interpolated elevation, a certain amount of planimetric shift must be allowed when testing for vertical accuracy.[4]

The quality of a map has also been judged by comparisons made on an areal basis rather than on the point basis just described. For instance, in a study of the effect of reduction of map scale (increase of generalization) on map accuracy, Salisbury and LaValle compared the local relief data as measured from the units of a square-mile grid which they laid over topographic maps of three different scales covering the same area.[5] Correlation coefficients were derived describing the correspondence between the local relief data of the maps of the largest scale (1:24,000) and the data as obtained from each of the smaller scales (1:62,500 and 1:250,000). These coefficients were employed to express the direction and strength of the effect of scale reduction on map accuracy. F. Uhorczak and M. Zdobnicka also used an areal basis for their comparative studies between choropleth and isarithmic maps.[6]

For the purposes of this study, the method of point checking is considered to be more suitable than that of area checking. An isarithmic surface, as pointed out previously, is assumed to represent a continuous undulating surface, and any delimitation of subdivisions of the surface will inevitably be arbitrary. Subdivisions are required, however, in order to evaluate map accuracy on an areal basis. To compare a resultant map against its original surface by area checking, one could, for example, compare the geographical means of arbitrary subareas on the original surface with their counterparts on the resultant map. But in the process of determining the mean values of the subareas from the maps, the positive and negative map errors (differences between the values on the original surface and the corresponding values on the resultant map) which occurred at points within the subareas would tend to cancel each other. Therefore the difference between the two means would not reveal the real magnitude of the errors.

Still another alternative in investigating the quality of a resultant isopleth map is to compare directly the total volume of the isopleth surface and that of the original one. The volumes may be calculated and their difference is assumed to be the map error. If we want to know where the errors occur, however, we again have to employ an area or point checking method. Each surface (original or isopleth) can also be described mathematically, for example, by a polynomial or Fourier series, and comparisons can be made of these terms. These would be generally less revealing.

Sampling Procedure

Because both the original distributions and the resultant isopleth maps are conceptually continuous surfaces, there are an infinite number of points on each; consequently a sampling procedure was adopted in order to select a finite number of points at which to examine the magnitudes and characteristics of the errors introduced by the isopleth mapping process. This process involves the selection of a sampling method, a sample size, the location of the sample points, and the derivation of the sampling discrepancies.[7]

There are two types of populations to be sampled: the four original distributions and the eighty resultant isopleth maps. The first provide samples of original values and the second, samples of map values. The two populations are obviously related, since the resultant maps are derived from the original distributions. Because the resultant maps are prepared from data obtained by way of twenty different combinations of unit area aggregation and hexagonal transformation, applied to each original surface, an original surface is associated with twenty resultant maps. Accordingly, many decisions concerning the sampling procedure are made with reference only to the original surfaces, on the assumption that these decisions would be properly applicable to the resultant maps as well.*

*Naturally, the planimetric positions of the sample points on the original maps must be identical with those on the resultant maps, since each original value is to be compared with its corresponding resultant map values in order to derive the map errors at a particular point on the various resultant maps.

The Sampling Method. The three-dimensional characteristics of the original surfaces, I–IV (Figures 4–7), were first examined from the point of view of possible sampling methods. Two things are apparent. First, with the exception of Surface I, the degree of surface variation in terms of both local relief and texture differs from place to place; that is, some parts of a surface have fewer variations and other parts more. Therefore, it is necessary to employ a method in which a sample of given size represents fairly well points at all levels of the spatial variation, for it seems reasonable to assume that higher surface variations would be likely to induce larger map errors. Second, for Surfaces II, III, and IV, where the variations are irregularly arranged, one may also assume that when these surfaces are subdivided, that is, statistically stratified, each subpopulation of a surface is likely to have a smaller variance than the variance for the whole population. In view of these two assumptions, one may assert that when sampling from a three-dimensional distribution, either a two-dimensional stratified random sampling or some kind of systematic sampling is likely to provide a better sample than would simple random sampling.[8]

For any three-dimensional distribution a careful study of its spatial variations must be carried out before one can determine whether or not systematic sampling is more effective than stratified random sampling. Such a study involves an analysis of the auto-correlations between the z values of a large number of points at different distance intervals in the population (the infinite z values of the surface), and this in turn requires the construction of a graph (called a correlogram) in which x is the distance between points and y is the correlation coefficient between points. For most geographical quantities the correlation (positive) between pairs of points would vary inversely with the distance between the points. If the plotted curve on the correlogram is concave upward, systematic sampling is most effective for that population. If the curve shows some other curvature, a stratified random sampling plan is probably superior.

Correlograms for Surfaces I and II would probably produce concave curves, and therefore systematic sampling might be superior for these two surfaces. Surfaces III and IV, on the other hand, are much more complex, and it is very likely that curves of correlograms for them would be more irregular and less uniformly concave upward. Since the present research is analogical, it is desirable to use only one sampling method for all surfaces, rather than four methods each developed individually for a surface. Accordingly, stratified random sampling is employed for all surfaces, with one random sample taken within each stratum.

The Sample Size. The populations sampled in this study are spatial distributions; therefore, the stratification of the data requires the establishment of some kind of system for subdividing the sample area. One sample is to be taken in each stratum; the number of strata is therefore also the size of the sample.

The consideration of a sample size involves analyzing samples of various numbers of points in terms of how well they reflect the frequency distribution of the z values of the population. On the original surfaces, of course, each point has its own z value, and theoretically one could plot an ordinary frequency distribution graph to show the variation in the infinite number of occurrences (on the y axis) as related to the magnitudes of z values (on the x axis). It is essential that a sample be large enough to represent properly the relative frequency of the z values of the population; that is, the frequency distribution curve of the sample should have the same characteristics as the curve derived from the total population. A considerable difference between these two frequency curves would be particularly critical if there were actually a significant correlation between the z values and the map errors.

In order to determine an optimum sample size which would represent fairly well the z frequencies, some trials are performed based on the data collected from Surface III. (We assume that a sample large enough to be functional for Surface III, which is fairly complex and has distinctive linear concentrations, would be adequate for the other surfaces.) Samples of 16, 24, 30, 40, 96, and 140 points are taken. First, the sampling z value frequency distributions are examined in terms of their means and variances, both of which are remarkably close among the samples. Next, a frequency distribution of z values is plotted for each sample. In order to compare the characteristics of these frequency distributions the number of occurrences, that is, the frequency of the observations, must be kept constant. Accordingly, all frequency distributions are converted to 96 points and plotted as comparable histograms (Figure 14). The actual frequency distribution of the total population is, of course, not known, since one cannot graph an infinite number of points; however, it can be approximated by very large samples. The samples of 96 and 140 points are used as references and the other samples are compared with them.

The small samples of 16 and 24 points are considered inadequate because they fail to include several occurrences of z values within the ranges that are presented by the two largest samples (Figure 14, A and B). On the other hand, the histogram of the sample of 30 points seems to be a fair representative of the frequency distribution (Figure 14C), and that sample size is also manageable in a practical sense. Since a single sample is only one of the infinite number of possible samples of that size, additional examples are needed before one can reasonably accept the sample size. Two more samples of 30 points were taken (Figure 14, D and E). The three samples of 30 provide a good representation of the z values falling within the range of 3 to 18, the range included by the largest sample. Moreover, they retain the general form of the histograms of the two largest samples, namely, a slightly positively skewed distribution. In short, a sample of 30 points seems to be a reasonable choice.[9]

The Location of the Sample Points. As was observed in Chapter 2, the positions of the isopleths near the margins of the resultant maps are less certain because of the breakdown of the network of the control points in these sections; thus map errors are likely to be larger there. This is a not uncommon consequence of unavoidable cartographic procedures, and it obviously affects the fidelity of a map. But this phenomenon is independent of the factors under investigation; furthermore, it is limited to a special area of the map. Therefore, samples of map errors taken near the margins of the resultant isopleth maps would probably be persistently relatively larger than samples taken from other parts of the maps, and this in effect would introduce a kind of bias error into the sampling procedure. In order to eliminate the possibility of such an error, an inner border, one inch from the original border, is drawn on all surfaces and resultant maps, and the samples are taken within these secondary borders. The sampling areas, therefore, are six inches by nine inches at the working scale. Each sampling area is divided equally into thirty subareas,

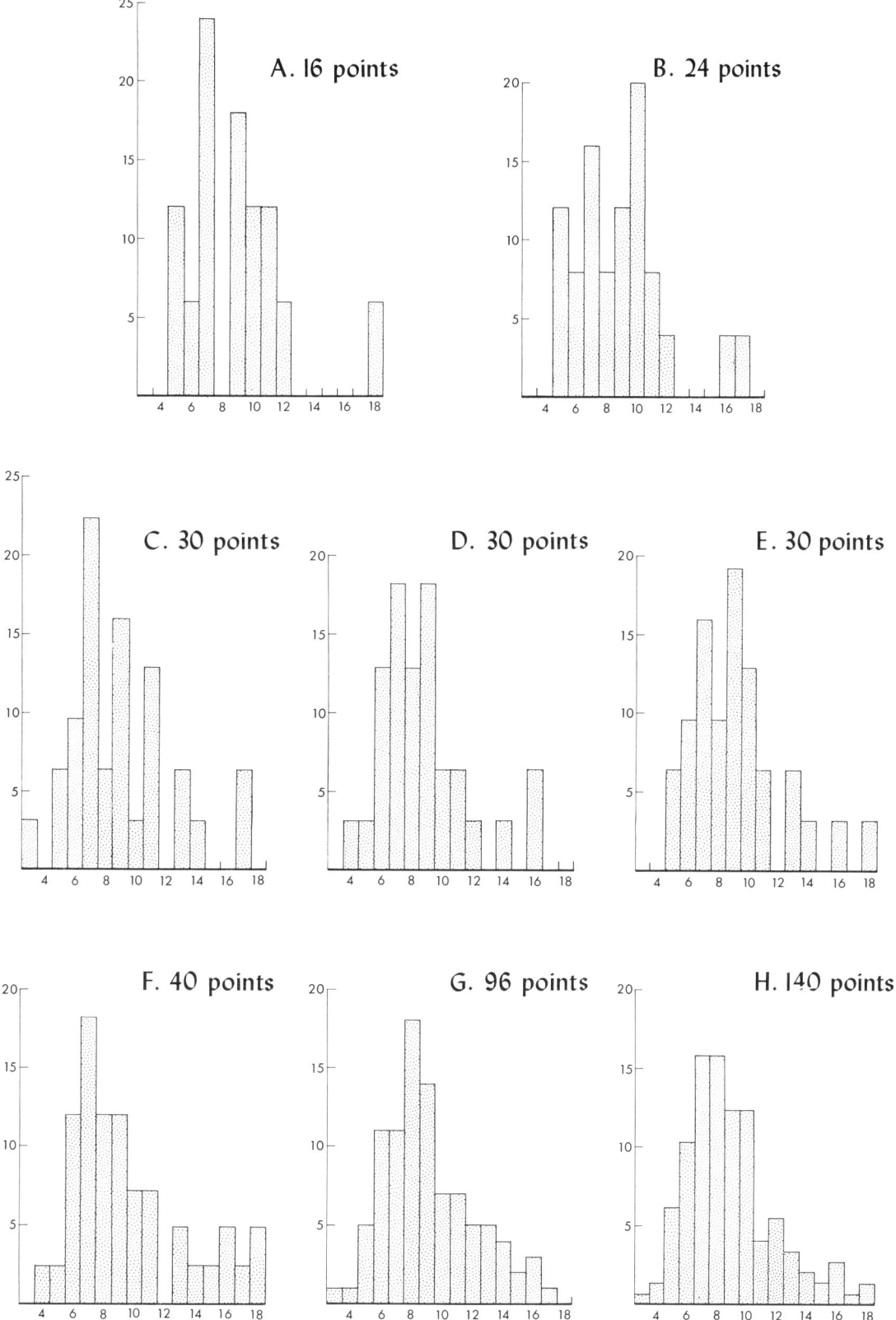

Figure 14. Frequency Distributions of z Values as Obtained from Various Sizes of Samples. The magnitudes of the z values are shown on the x axis, and their frequency on the y axis.

Plates 1–16

THE RESULTANT MAPS

The following sixteen plates show the resultant isopleth maps, in black, in combination with the respective original surfaces, in brown, from which they were derived. The code on each resultant map designates the original surface, the pattern of unit areas, and the size of the hexagon employed in the transformation. For example, SI-PI-HA refers to Surface I, Pattern I, and Hexagon A (the smallest hexagon size). The resultant maps show only the isopleths and the isarithms of the original surfaces. The unit area patterns are shown, in combination with the original surface, in the lower right of each plate.

PLATE 1

PLATE 2

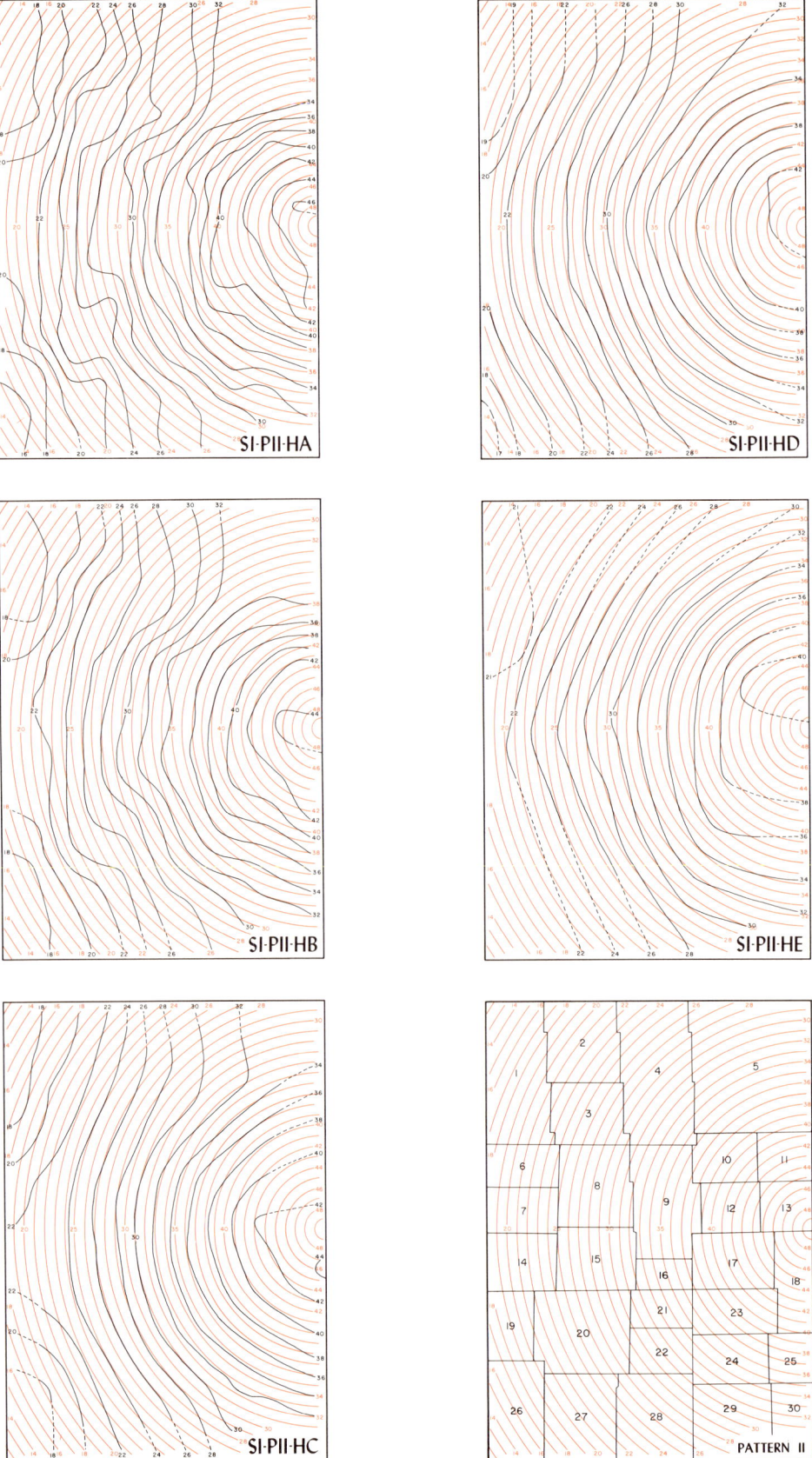

PLATE 3

PLATE 4

PLATE 5

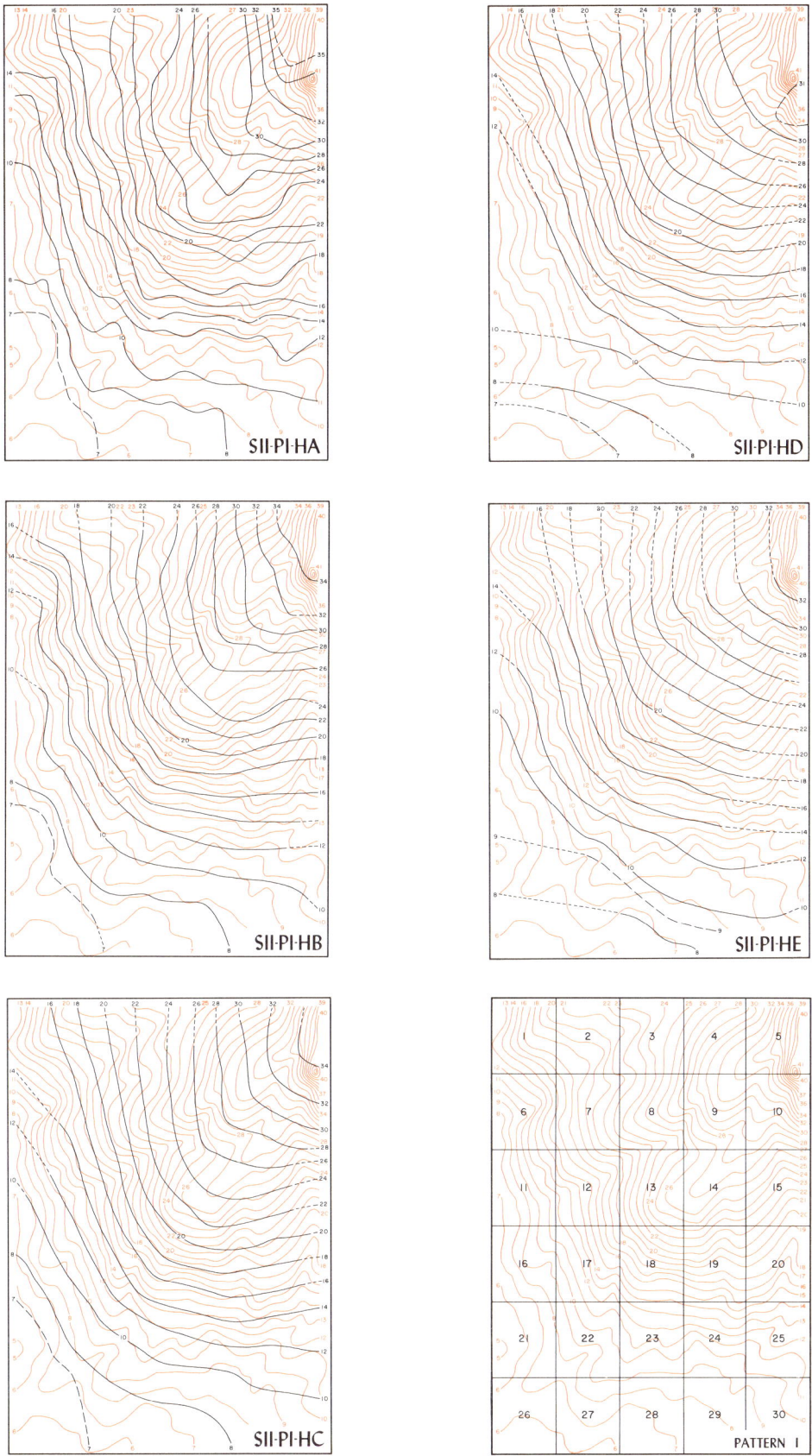

PLATE 6

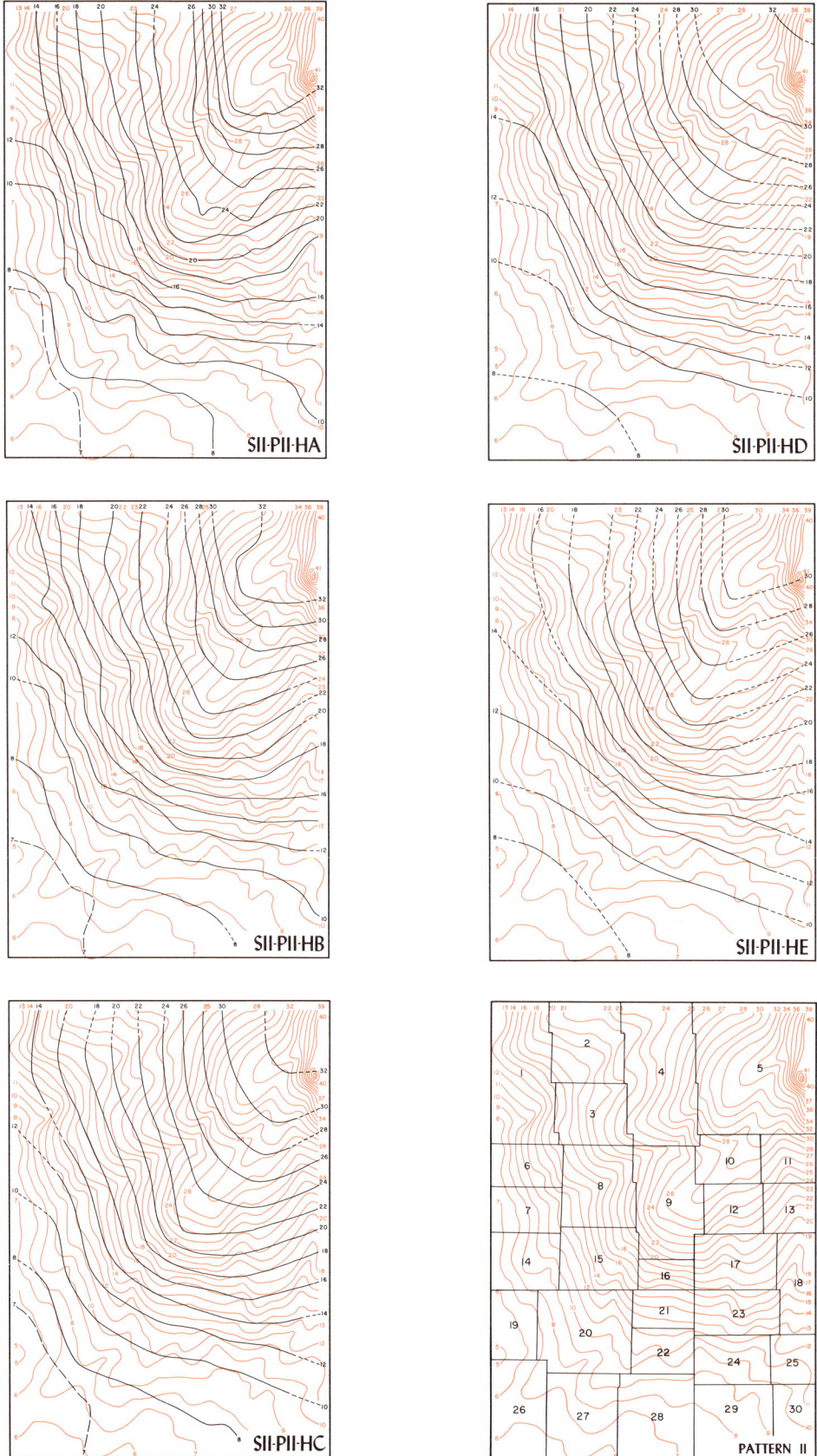

PLATE 7

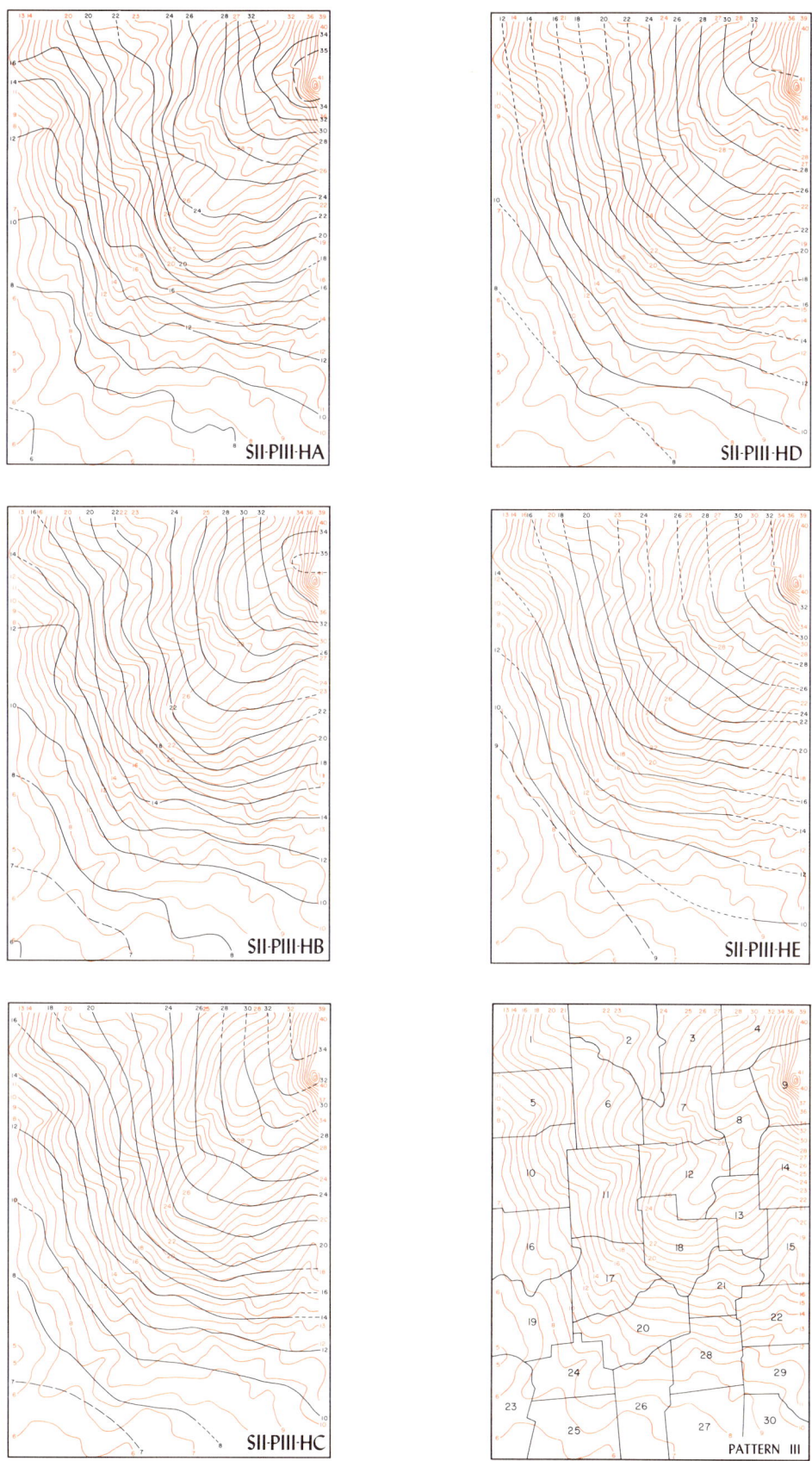

PLATE 8

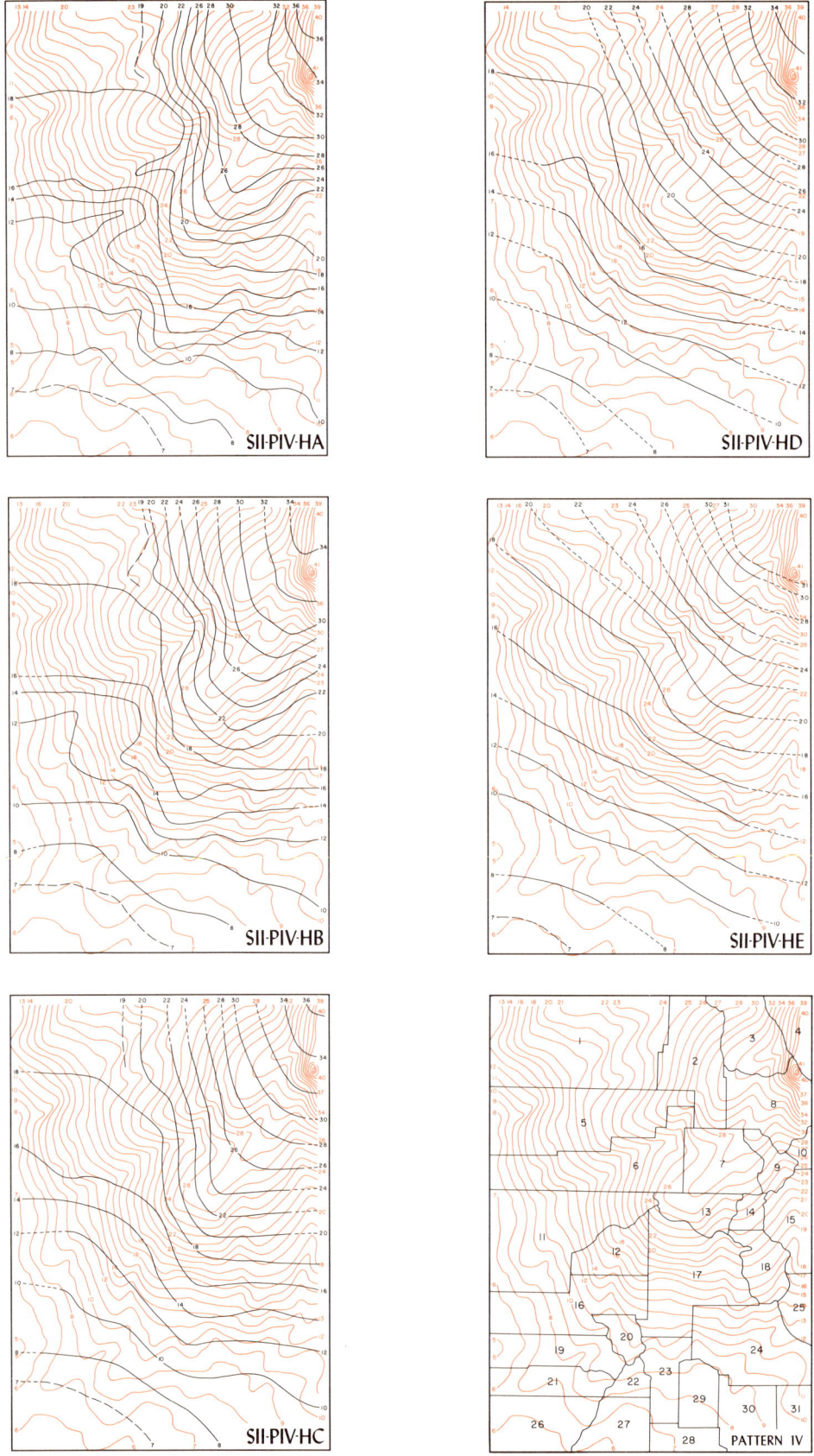

PLATE 9

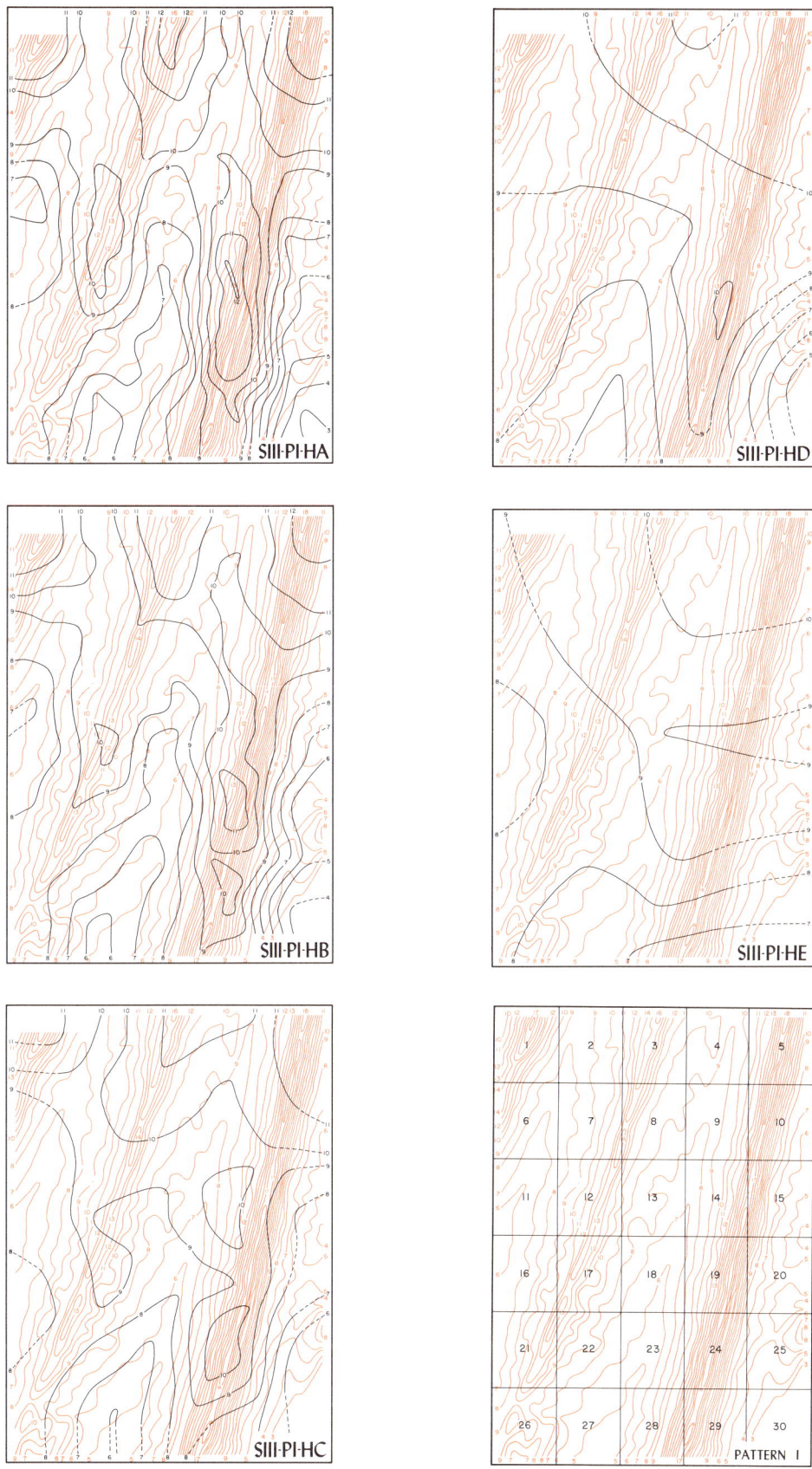

41

PLATE 10

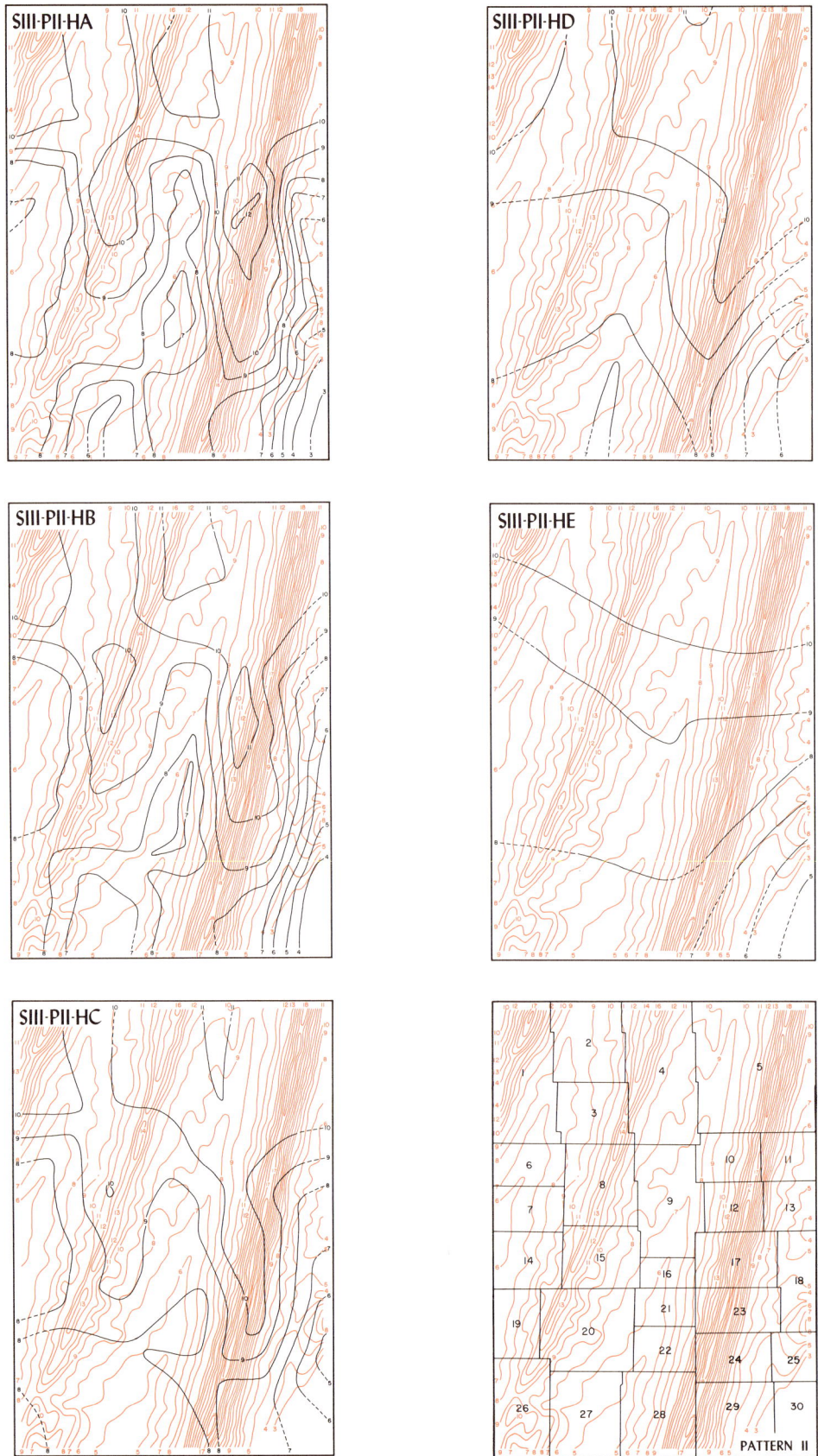

PLATE 11

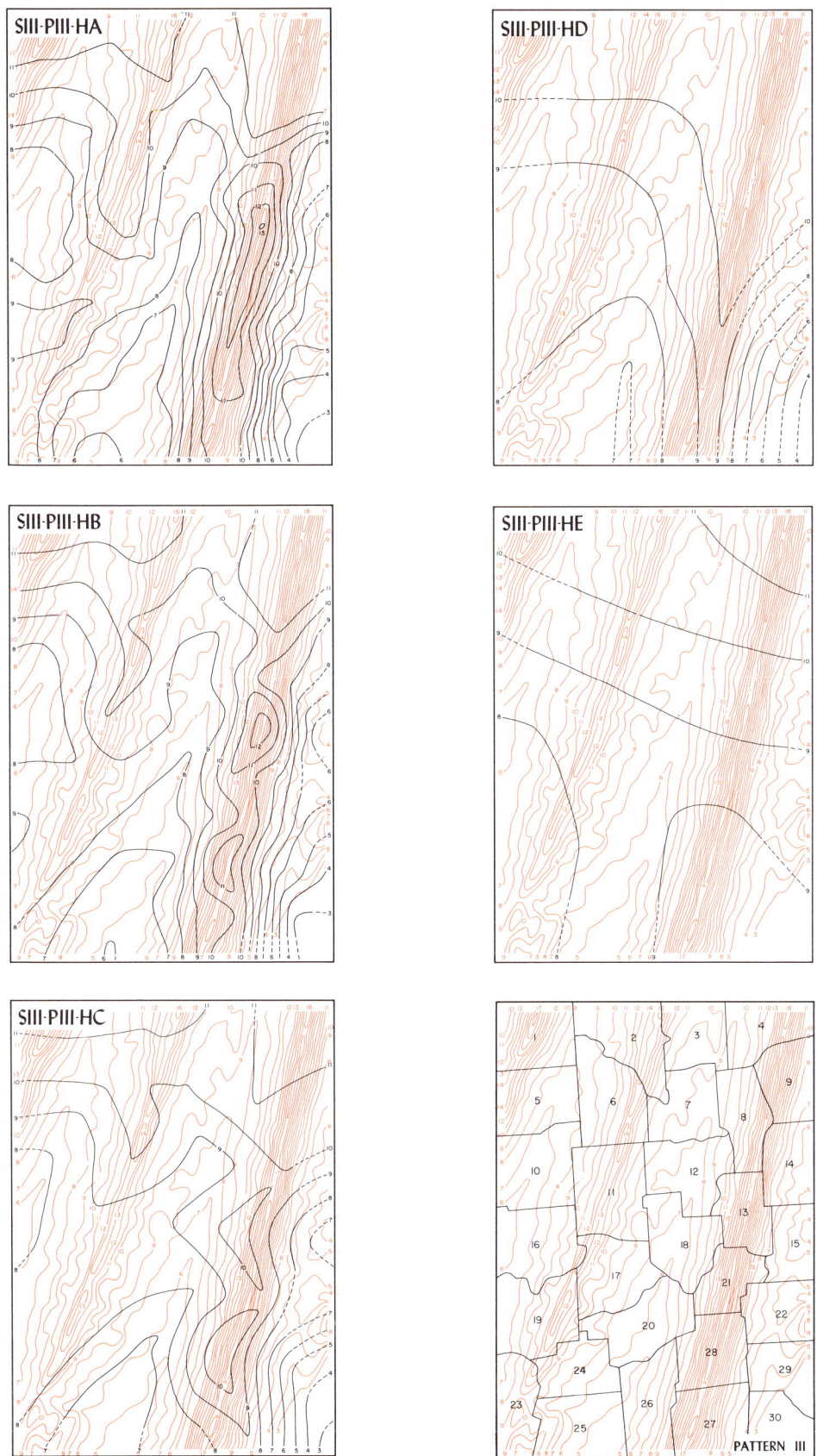

PLATE 12

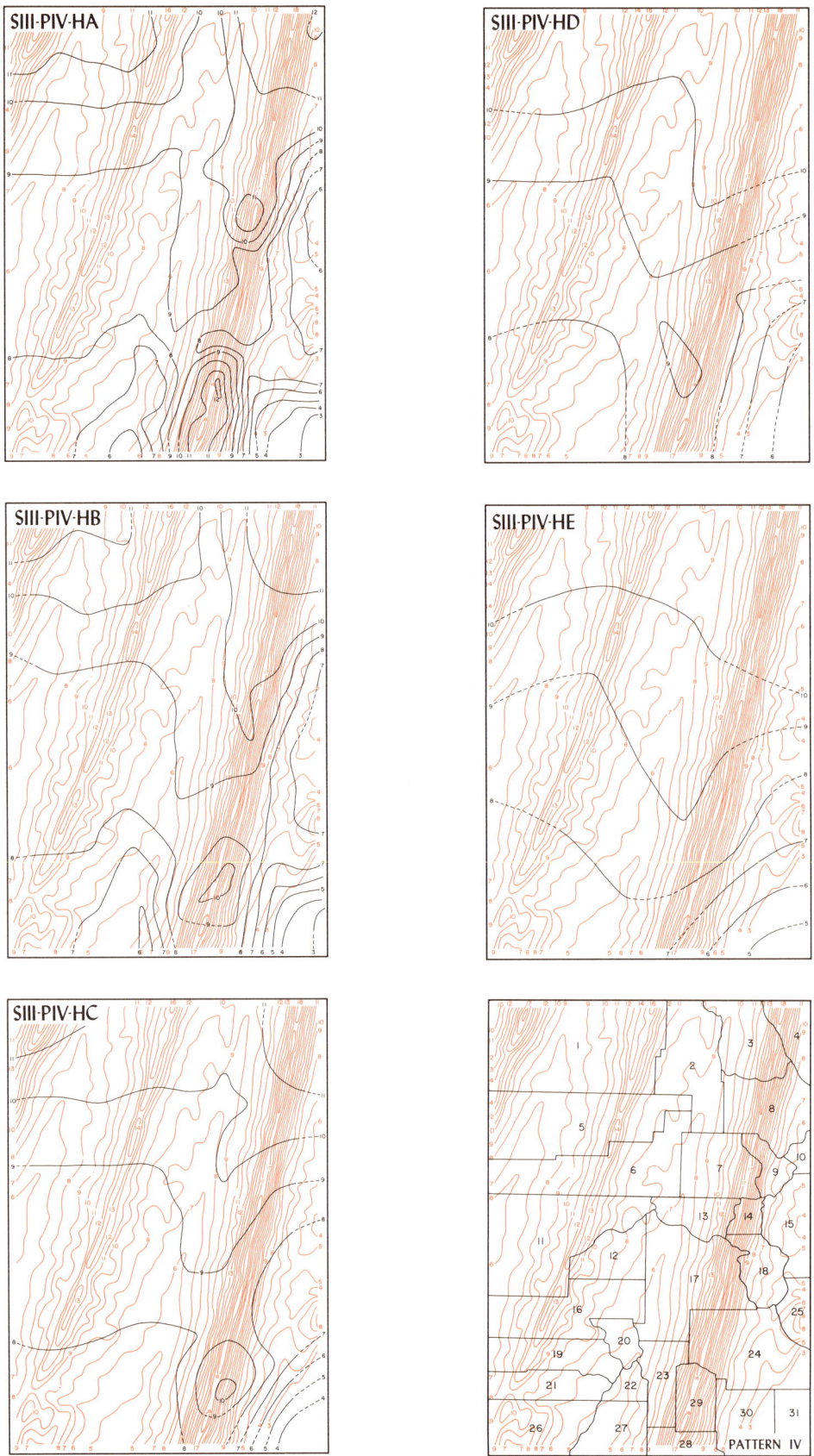

PLATE 13

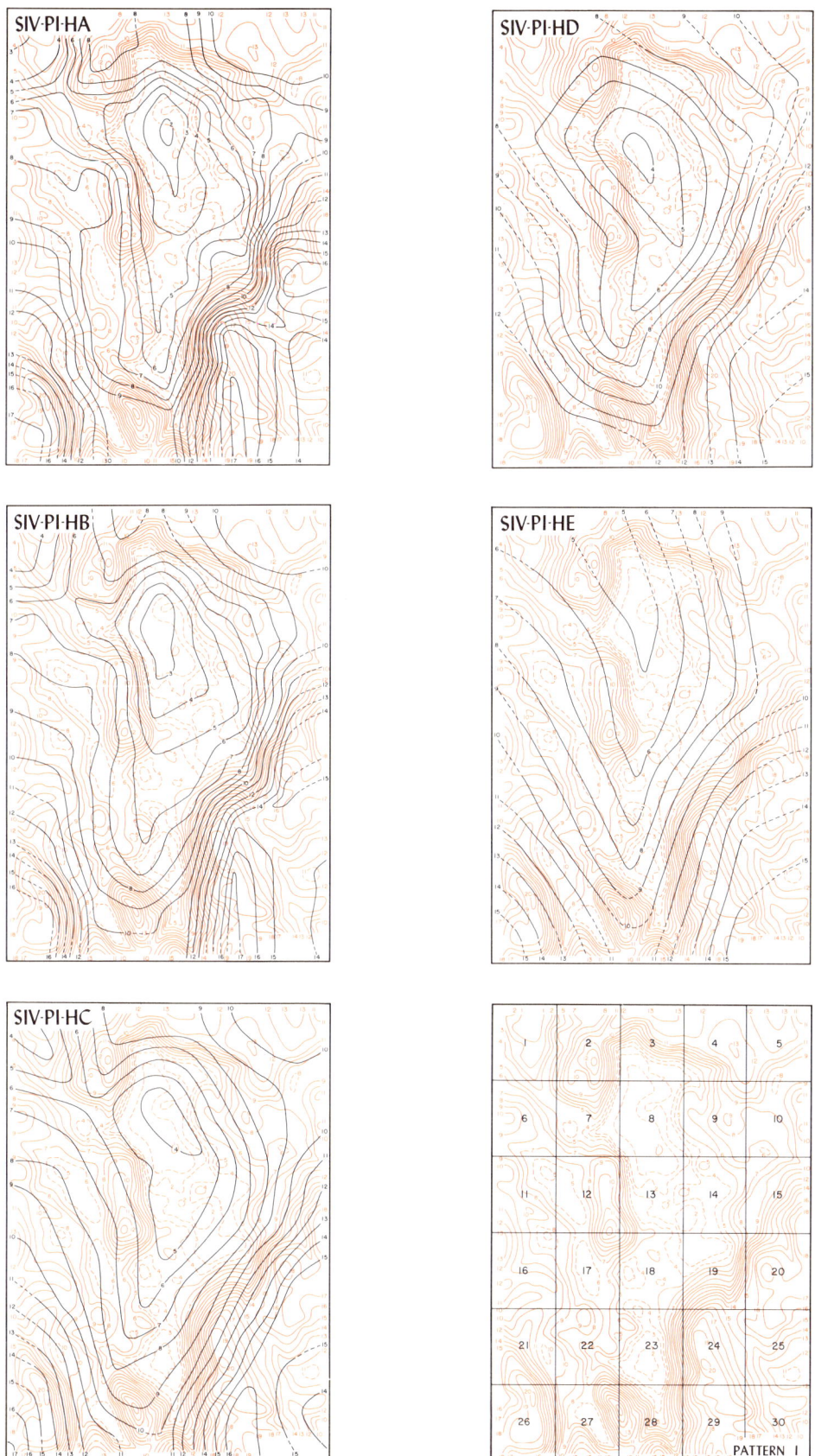

PLATE 14

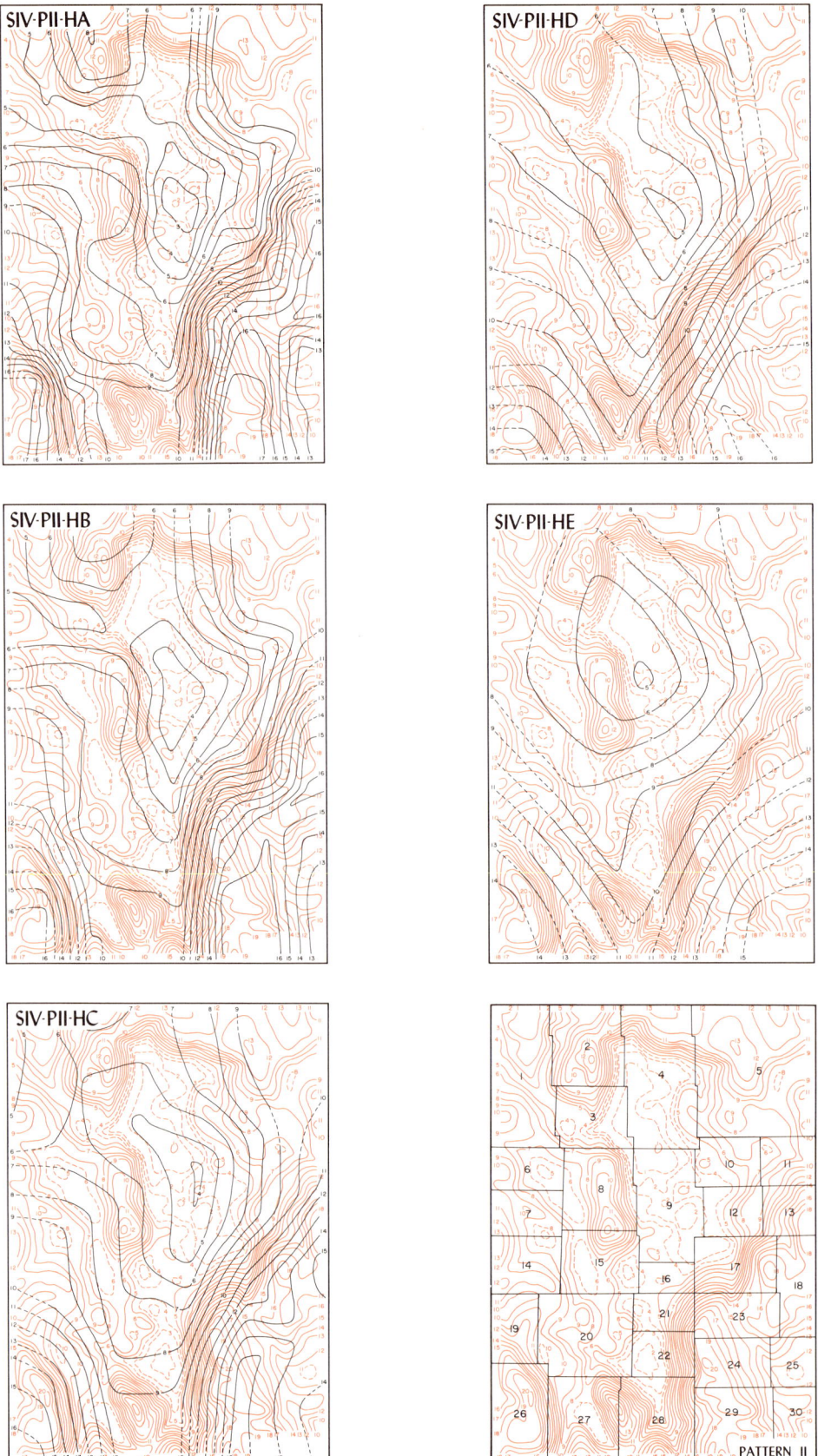

PLATE 15

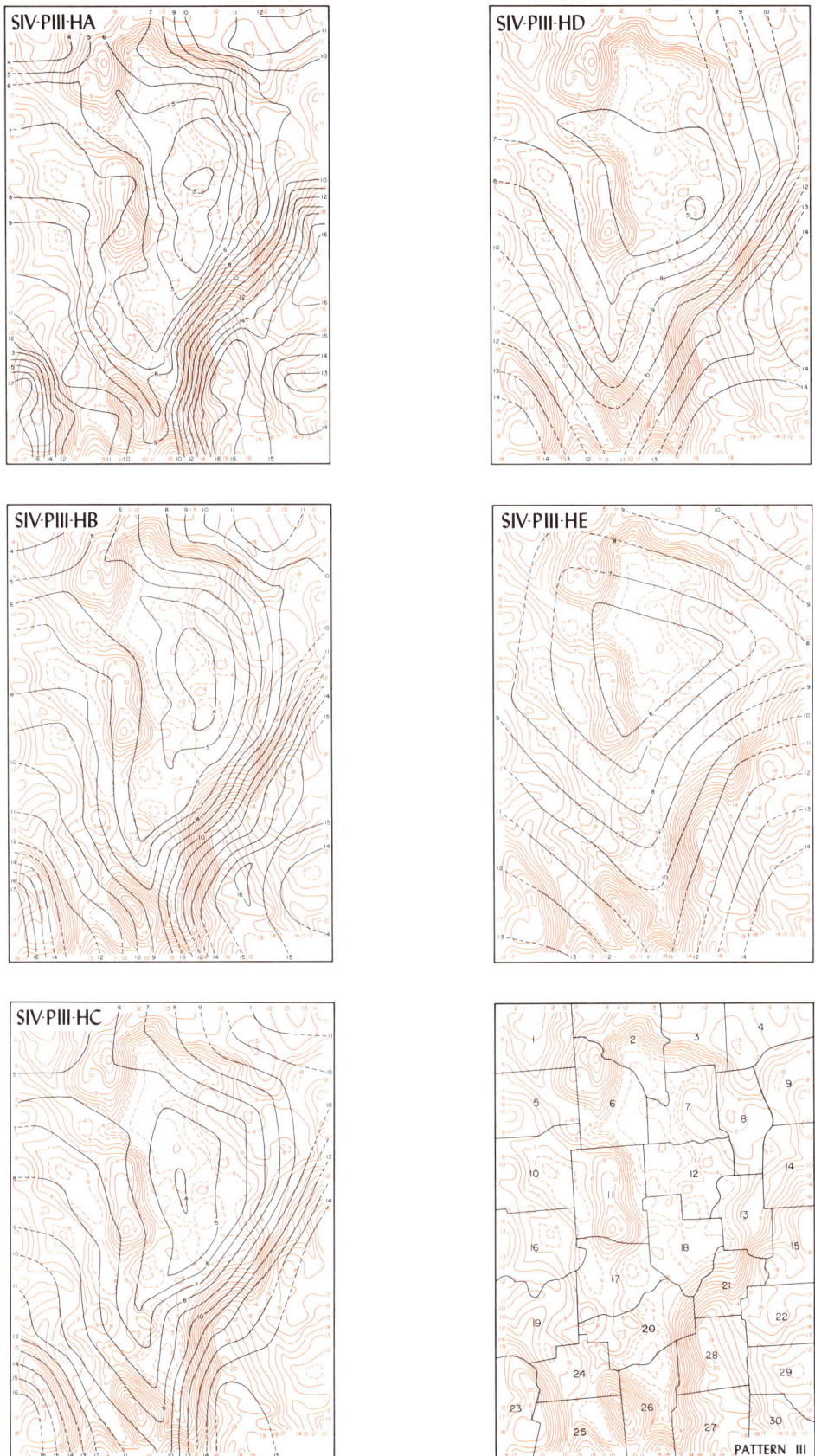

PLATE 16

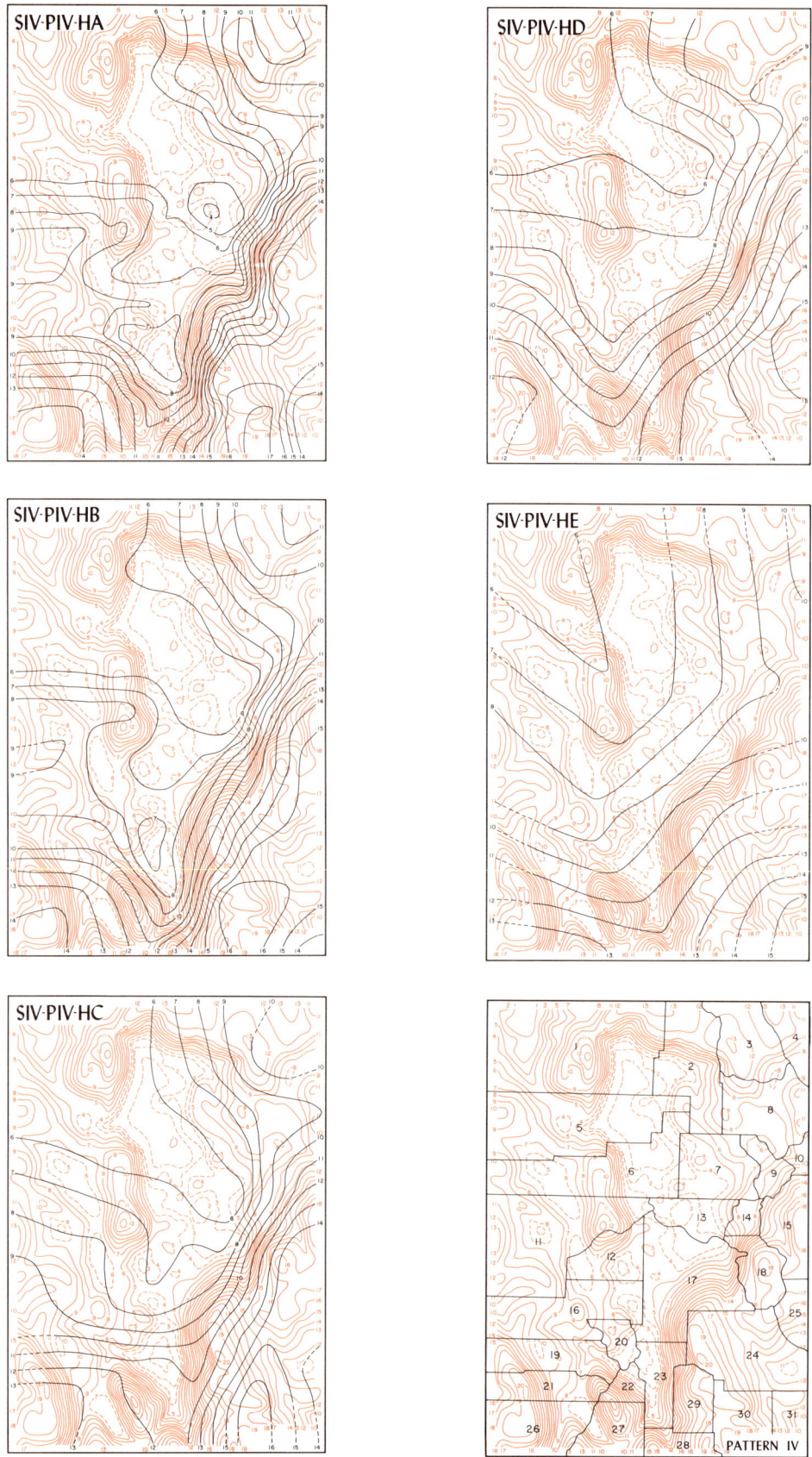

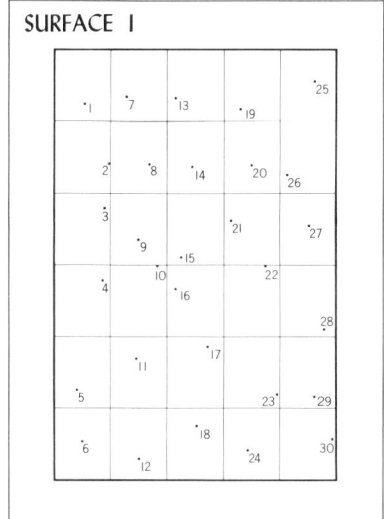
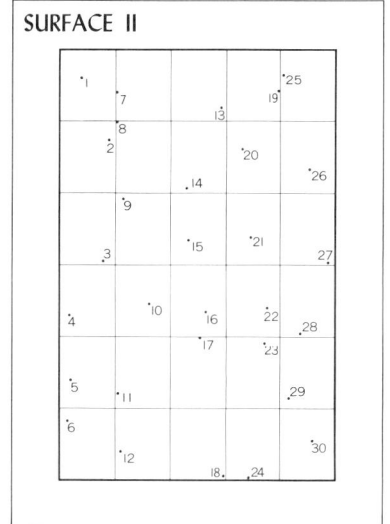
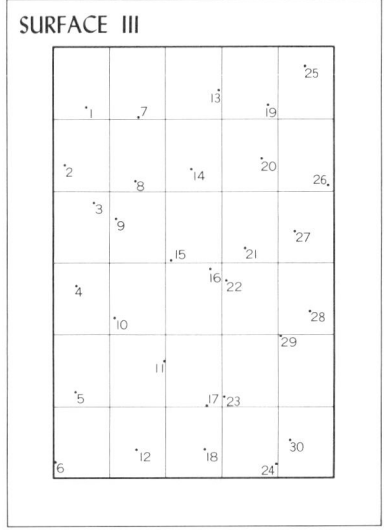
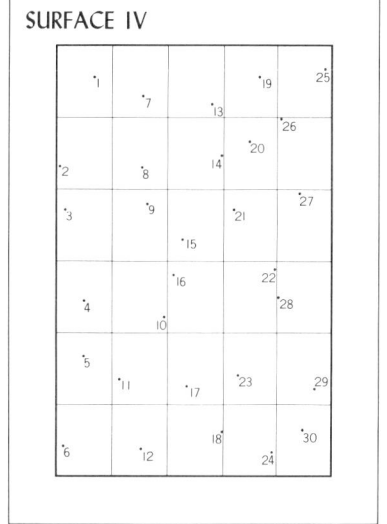

Figure 15. Locations of Sample Points on the Four Original Surfaces. (The same locations were used in connection with the set of twenty isopleth maps derived from each surface.)

within each of which a point is randomly positioned (Figure 15).

The Derivation of Sample Values and Map Discrepancies. The z values at the sample points for each of the four original surfaces plus the eighty resultant maps are obtained through linear interpolation from the isarithms. These values are then converted to quantities representing map discrepancies.

The map error at any point is defined as the discrepancy (D) between the value (G) on an original surface and the map value (M) at the corresponding point on an associated resultant isopleth map. Thus, for each surface, the map errors may be expressed as $D_{ij} = M_{ij} - G_j$, where $i = 1, 2, \ldots, 20$ (the sequence of the isopleth maps derived from one surface, for example, SI-PI-HA, SI-PI-HB, ..., SI-PIV-HE), and $j = 1$,

49

2, ..., 30 (the numerical order of the sample point). Since for one surface the G's do not vary from $i = 1, 2, \ldots, 20$ but do vary from $j = 1, 2, \ldots, 30$ — that is, they do vary from one sample point to another — a set of G's for one surface is to be compared with each of the 20 sets of M's associated with it.

In summary, for each of the four original surfaces, there are 30 G's, 600 M's, and, therefore, 600 D's. These quantities form the basic data for the statistical analysis of the fidelity with which the isopleth maps portray the original distributions.

Level of Generalization

Preliminary analysis of the relations among the magnitudes of the D values for a given sample point of a given original surface (that is, the discrepancy between the value at a sample point on the original surface and the values at that sample point on each of the twenty resultant maps) indicates that their magnitudes are reasonably constant; that is, some sample points have consistently large discrepancies and some have small ones. This seems to indicate that the magnitudes of the "average" discrepancy at some of the sample points are much greater than at others. Further examination reveals something different, however. For example, for the twenty isopleth maps derived from Surface III, the D's at sample point 1 vary from 1.8 to 2.8; at point 5 they vary from -2.7 to -3.9; and at point 6, they vary from -0.7 to 0. Consequently one may say that the variation from point to point among the thirty D values for a single resultant map is considerably larger than the variation among the twenty individual discrepancies at a particular sample point on the twenty resultant maps.

This phenomenon occurs on all the maps but in varying degree; it is very pronounced on the isopleth maps derived from Surfaces III and IV and is least in those derived from Surface I. Clearly it is a consequence of some variable in the mapping process that markedly affects the character of the resultant maps; but on the other hand its occurrence seems entirely unrelated to the variations in unit area patterns (size and regularity) or to the sizes of the hexagons used in the transformation, these being (along with the characters of the surfaces) the elements affecting the accuracy of isopleth maps that are under study. This means that this variable has somehow to be included in the analysis.

The existence of such a variable is partly a result of generalization, a ubiquitous process in cartography. In this study, generalization is introduced at a particular level by the selection of the number of subareas in each unit area pattern; thirty are chosen as part of the experimental design.[10] The four unit area patterns, each with its array of subareas (enumeration districts), are the bases upon which the original distributions are transformed from continuous surfaces with an infinite number of point values to sampling surfaces, each with thirty aggregated areal averages. This establishes a given degree of generalization in the mapping procedure; no matter what combination of unit area and hexagonal patterns is used, a resultant map cannot portray the surface from which it is derived with a greater degree of fidelity than is permitted by this level of generalization. In other words, if each of the unit area patterns had been divided into sixty subareas instead of thirty, then each original distribution would have been represented by sixty aggregated areal averages, and all the isopleth maps would therefore have been correspondingly less generalized. This aspect of generalization is held constant in the experimental design by using essentially the same number of aggregated areal averages to represent the original surfaces, regardless of the combination of unit area pattern and original surface.

Unfortunately the basic generalizing process did not result in an overall uniform level of generalization since another factor entered the picture. Although this lack of uniformity is independent of the variables on which the study focuses, it clearly does affect the fidelity of the isopleth maps. It comes about as follows. The chance location of the unit areas causes some of the detailed features and some of the high or low values on an original surface to be reduced or lost as a result of the generalizing process; consequently, the resultant maps with isopleths based on averaged values portray surfaces that are much smoother than the original distributions. Consequently, if the location of a sample point on an original surface were such that its value by chance happened to be either much above or much below the generalized values of the resultant maps, then the D values at this point would be consistently large.

Figure 16 illustrates how this could occur. On Surface III points 5 and 6 are both located at the lower end of a linear concentration which is largely lost in the isopleth mapping process. Instead, the resultant maps broadly present the area as being one of little gradient

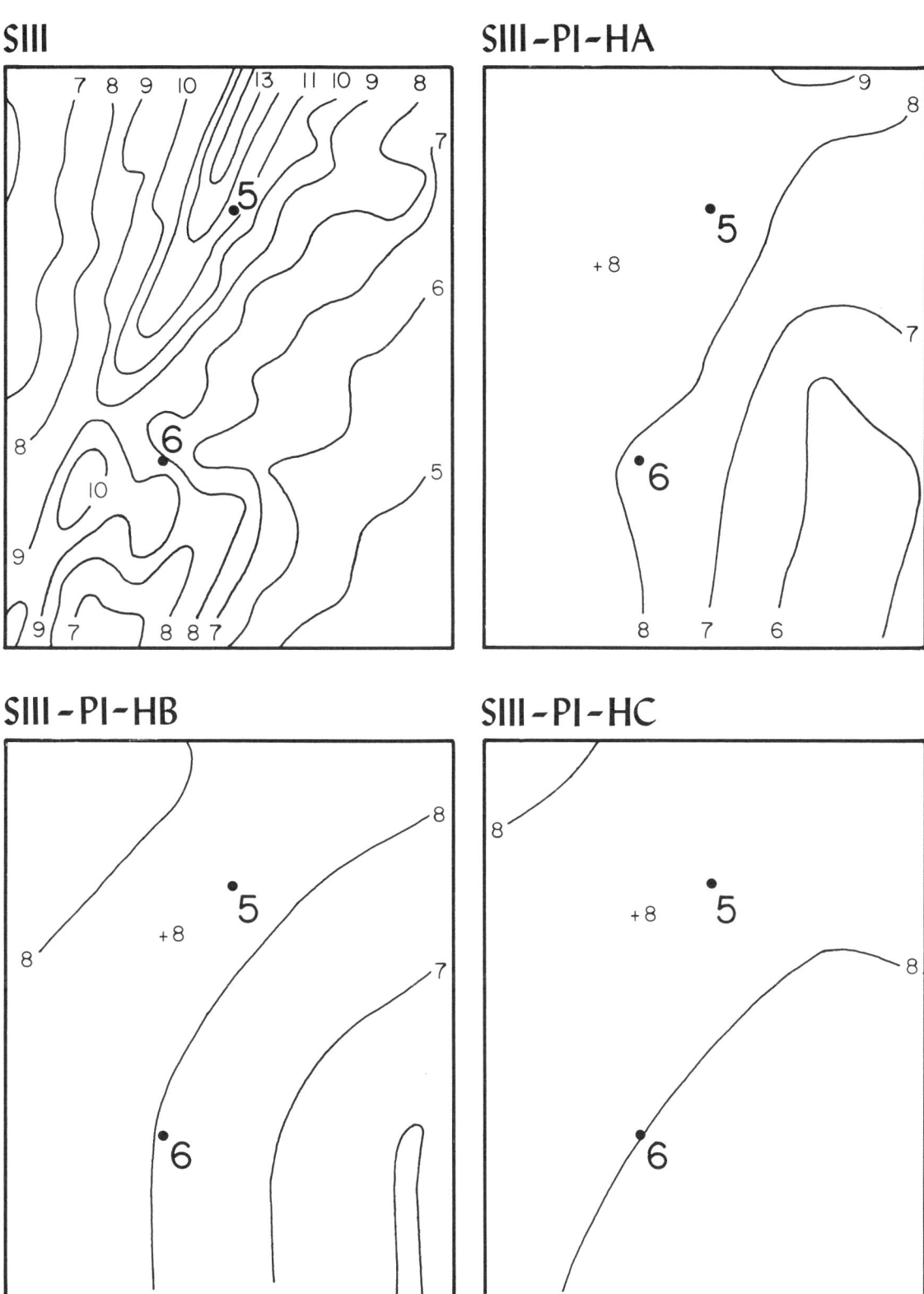

Figure 16. Effects of the Chance Locations of Sample Points on Magnitudes of D Values in Relation to Generalized Average Values. Illustrated here are the lower left corner of Surface III and three of the resultant isopleth maps showing sample points 5 and 6.

with isopleth values of 6, 7, 8, and 9. The value of the original surface at point 6 is 8.1 and therefore the D values (difference between original surface and resultant map) recorded for this point are uniformly small. On the other hand, point 5, with an original value of 11.4, is high compared to any unit area average that would be derived for this area; thus it generates large D values. Had point 5 been located a little to the right on the original surface, it would have had an original value near 8 or 9, and the D values would therefore have been consistently smaller.

In order to take account of the occurrence of these variations, it is necessary to quantify them in some manner. To obtain an objective value which can be employed in the analysis to represent the variation in a consistent way for all patterns, a circle (with an area equal to that of the average size of the unit areas) is centered at the location of each sample point on each of the four original surfaces. For each such circle an average z value (C) for the enclosed area is determined by the same method used to derive the unit area values. The circle thus represents the average-size "filter" used in the generalization process.[11] The quantity C is therefore approximately the value that would be obtained were an average-size unit area to occur with the sample point location at its center. The procedure of introducing the C values is based on the assumption that the value of M (z value at a sample point on an isopleth map) is likely to vary consistently with the value of C. If the value of a sample point on an original surface (G) is similar to the average determined for the circle (C), then the D values (that is, $M - G$) for that point would generally be small.

From the values of C and G the quantity E_j is calculated:

$$E_j = C_j - G_j \qquad j = 1, 2, \ldots, 30$$

where j is the numerical order of a sample point. The E values, which are entirely derived from an original surface, are, of course, identical throughout $i = 1, 2, \ldots, 20$, this being the sequence of the resultant isopleth maps derived from that surface. If the variations induced by sample location do affect the magnitudes of the D values as hypothesized, there should be a positive correlation between E_j and the sampling discrepancies, D_{ij}. To test this, a simple linear correlation was run between the absolute values of E_j and D_{ij} for the six hundred observations of each surface. The results, as summarized in Table 3, are indeed revealing

Table 3. Correlation between D and E Values

Surface	Correlation Coefficient
I	0.32
II	0.55
III	0.81
IV	0.72

and indicate a strong association between the two variables, particularly for the complex Surfaces III and IV. The relatively low correlation coefficient for Surface I in Table 3 is also worth noting: it reflects the validity of a standing cartographic assumption that for a simple surface the number of unit areas, the density of control points, and the level of generalization are much less critical to map accuracy than for a complex surface.[12]

For the purposes of this study the variations resulting from the chance sample point location are an unwanted factor. Since this investigation focused major attention upon the effects of differences in surface complexity, in unit area size and shape, and in the size of hexagonal transformation, the "built in" level of generalization is automatically included. It is possible, fortunately, to include the unwanted factor of sample point location in part of the statistical analysis in such a way as to separate the variation caused by it from that caused by the other variables. In other analyses, however, it remains one of the causes of variation. It is by no means an overriding factor.

4

Statistical Analyses of the Fidelity of the Isopleth Maps

THE data consisting of the sample discrepancies which were collected for each of the eighty resultant maps derived from various combinations of surfaces, unit area patterns, and hexagon sizes may be subjected to a variety of statistical analyses. In view of the experimental design the most suitable general approach is the technique called the analysis of variance (ANOVA). In addition one may usefully examine the absolute magnitudes of the errors in the resultant maps, the dispersions of these errors, and the algebraic values of the errors.

The Analysis of Variance

As employed here, the ANOVA tests the hypothetical assumption that the selection of alternative combinations of unit area patterns and hexagon sizes does not affect the accuracy of the resultant maps — that is, that the isopleth maps do not vary in "fundamental" accuracy and that any differences among them are solely a result of sampling errors. In essence the ANOVA technique measures the total variation in the data and separates it into two parts: the part attributable to the group of causes which are being examined in the analysis, and the part which must be ascribed to other causes. In addition the partial variation due to each individual cause is also identified.

The D values to be analyzed (the differences at the sample points between the original surfaces and the isopleth maps) naturally consist of both positive and negative values. To some extent these values cancel out upon summation, and consequently the mean of a group of discrepancies is reduced. In mapping, however, such measures are all errors, regardless of sign, and the real magnitude of the consequences of the isopleth process can be analyzed only by ignoring the signs of the D values in the ANOVA.

Since the four original surfaces are not quantitatively comparable, the most useful approach is to carry out a separate ANOVA for each surface. There are six hundred observations in each analysis. The unit area patterns were grouped into a 2×2 matrix on the basis of (1) regular shape (PI and PII) or irregular shape (PIII and PIV) and (2) uniform size (PI and PIII) or variable size (PII and PIV). The sizes of the hexagons provided five classes. The variations induced by the chance locations of the sample points are included simply by the ordinal numbers of the sample points, of which there are thirty.

The four-way ANOVA model postulated to explain the composition of the total variation of the D values is

$$y_{ijkm} = \mu + P_i + S_j + H_k + L_m + PS_{ij} + PH_{ik} + PL_{im} + SH_{jk} + SL_{jm} + HL_{km} + PSH_{ijk} + PSL_{ijm} + PHL_{ikm} + SHL_{jkm} + E_{ijkm}$$

where $i = 1, 2,$ $k = 1, 2, \ldots, 5,$
$j = 1, 2,$ $m = 1, 2, \ldots, 30.$

The model attributes the total variation to the general mean of the data (μ), the variation due to the difference of unit area shape (P), the variation due to the difference of unit area size (S), the variation due to the difference of hexagon size (H), the variation due to the difference of the sample point location (L), and the two-way, three-way, and four-way interaction effects among P, S, H, and L factors. In the model ($2 \times 2 \times 5 \times 30$) there are no replicates within the cells, that is, there is only one observation in each cell. Therefore, the error term E_{ijkm} is the sum of the four-way interaction $PSHL_{ijkm}$ and the residuals of the ANOVA. The model is designated as Model I by the statisticians.[1]

The summary of the F tests and the complete four-way ANOVA tables are shown in Tables 4 to 8. For all surfaces the main effects — that is, the shapes of the unit areas (P), the sizes of the unit areas (S), the sizes of the hexagons (H), and the sample locations (L) — are all statistically significant at the 5 percent level. Also significant are the three two-way interactions (PL, SL, HL) and one three-way interaction (PSL) between sample location and various other effects. Although these F values are all significant, it is evident from the ANOVA tables that they vary a great deal. For instance, the effect of unit area shape (P) is significant for all four sur-

Table 4. Results of *F* Tests of Four-Way ANOVA
Using Absolute Values

Source of Variation	Significant at 5% Level				Significant at 1% Level			
	SI	SII	SIII	SIV	SI	SII	SIII	SIV
P	x	x	x	x	x	x	x	x
S	x	x	x	x	x	x		x
H	x	x	x	x	x	x	x	x
L	x	x	x	x	x	x	x	x
PS	x	x			x	x		
PH		x						
PL	x	x	x	x	x	x	x	x
SH			x					
SL	x	x	x	x	x	x	x	x
HL	x	x	x	x	x	x	x	x
PSH								
PSL	x	x	x	x	x	x	x	x
PHL								
SHL								

NOTE. When the required degrees of freedom were not available from the source tables, the closest value was used; e.g., the *F* value of 200 d.f. was substituted for that of 116. Source: E. S. Pearson and H. O. Hartley, eds., *Biometrika Tables for Statisticians*, vol. 1 (Cambridge: Cambridge University Press, 1958).

Table 5. Four-Way ANOVA Table of Surface I
Using Absolute Values

Source of Variation	Sum of Squares	Degrees of Freedom	Mean Squares	*F* Test
P	37.101	1	37.101	$F_{1,116} = 202.85$
S	42.987	1	42.987	$F_{1,116} = 235.03$
H	13.632	4	3.408	$F_{4,116} = 18.63$
L	124.957	29	4.309	$F_{29,116} = 23.56$
PS	12.848	1	12.848	$F_{1,116} = 70.25$
PH	1.223	4	0.306	$F_{4,116} = 1.67$
PL	33.024	29	1.139	$F_{29,116} = 6.23$
SH	0.973	4	0.243	$F_{4,116} = 1.33$
SL	45.912	29	1.583	$F_{29,116} = 8.66$
HL	83.337	116	0.718	$F_{116,116} = 3.93$
PSH	0.343	4	0.086	$F < 1$
PSL	27.647	29	0.953	$F_{29,116} = 5.21$
PHL	26.662	116	0.230	$F_{116,116} = 1.26$
SHL	24.218	116	0.208	$F_{116,116} = 1.14$
Error	21.212	116	0.183	
Corrected total	496.076	599		
Mean	637.364	1		

faces but *F* equals 203 for Surface I and 16 for Surface III (Tables 5 and 7).

Since the four original surfaces cannot be compared quantitatively, any comparison among the surfaces based on the relative magnitudes of mean squares and *F* values can be made only in very general terms. On Surface I the shape and size of the unit areas and the interaction between the two seem to be the major variables; size is a little more influential than shape. In the case of Surface II the shape and its interaction with the size of the unit areas are by far the most important factors. On Surfaces III and IV, however, the most impor-

Table 6. Four-Way ANOVA Table of Surface II
Using Absolute Values

Sources of Variation	Sum of Squares	Degrees of Freedom	Mean Squares	*F* Test
P	34.752	1	34.752	$F_{1,116} = 248.41$
S	10.507	1	10.507	$F_{1,116} = 75.11$
H	36.536	4	9.134	$F_{4,116} = 65.29$
L	403.387	29	13.910	$F_{29,116} = 99.43$
PS	29.837	1	29.837	$F_{1,116} = 213.28$
PH	1.819	4	0.455	$F_{4,116} = 3.25$
PL	34.830	29	1.201	$F_{29,116} = 8.58$
SH	0.698	4	0.174	$F_{4,116} = 1.24$
SL	44.277	29	1.527	$F_{29,116} = 10.91$
HL	62.034	116	0.535	$F_{116,116} = 3.82$
PSH	0.418	4	0.104	$F < 1$
PSL	41.633	29	1.436	$F_{29,116} = 10.26$
PHL	17.919	116	0.154	$F_{116,116} = 1.10$
SHL	17.848	116	0.154	$F_{116,116} = 1.10$
Error	16.232	116	0.140	
Corrected total	752.727	599		
Mean	719.853	1		

Table 7. Four-Way ANOVA Table of Surface III
Using Absolute Values

Source of Variation	Sum of Squares	Degrees of Freedom	Mean Squares	*F* Test
P	2.220	1	2.220	$F_{1,116} = 16.42$
S	0.742	1	0.742	$F_{1,116} = 5.49$
H	6.412	4	1.603	$F_{4,116} = 11.86$
L	1502.731	29	51.818	$F_{29,116} = 383.21$
PS	0.487	1	0.487	$F_{1,116} = 3.60$
PH	0.089	4	0.022	$F < 1$
PL	17.837	29	0.615	$F_{29,116} = 4.55$
SH	1.749	4	0.437	$F_{4,116} = 3.23$
SL	18.159	29	0.626	$F_{29,116} = 4.63$
HL	45.858	116	0.395	$F_{116,116} = 2.92$
PSH	0.904	4	0.226	$F_{4,116} = 1.67$
PSL	13.090	29	0.451	$F_{29,116} = 3.34$
PHL	9.541	116	0.082	$F < 1$
SHL	16.157	116	0.139	$F_{116,116} = 1.03$
Error	15.686	116	0.135	
Corrected total	1651.664	599		
Mean	1704.546	1		

STATISTICAL ANALYSES

Table 8. Four-Way ANOVA Table of Surface IV Using Absolute Values

Source of Variation	Sum of Squares	Degrees of Freedom	Mean Squares	F Test	
P	4.824	1	4.824	$F_{1,116}$	= 21.63
S	2.693	1	2.693	$F_{1,116}$	= 12.08
H	53.548	4	13.387	$F_{4,116}$	= 60.03
L	1913.580	29	65.986	$F_{29,116}$	= 295.90
PS	0.123	1	0.123	$F < 1$	
PH	0.291	4	0.073	$F < 1$	
PL	33.954	29	1.171	$F_{29,116}$	= 5.25
SH	0.078	4	0.020	$F < 1$	
SL	40.272	29	1.389	$F_{29,116}$	= 6.23
HL	111.158	116	0.958	$F_{116,116}$	= 4.30
PSH	0.801	4	0.200	$F < 1$	
PSL	13.490	29	0.465	$F_{29,116}$	= 2.09
PHL	21.859	116	0.188	$F < 1$	
SHL	27.842	116	0.240	$F_{116,116}$	= 1.08
Error	25.871	116	0.223		
Corrected total	2250.381	599			
Mean	3297.539	1			

tant variable is sample point location. The hexagon size follows as a poor second in the case of Surface IV. It is no surprise to find that the effect of sample point location is much more critical with respect to a complex surface than it is to a simple one. Another conclusion, however, is not so readily understood: only on the simplest original surface, Surface I, is the size of the unit area more important than the shape in contributing to the variations among the resulting maps. For the other three surfaces the shape is more critical. This curious finding will be examined in Chapter 5.

The analysis of group means — that is, the means of the main effects of the ANOVA — sometimes results in additional information and therefore such an analysis was carried out. No significant knowledge was gained from it, however.[2] Usually a set of assumptions is attached to a statistical technique, and the validity and usefulness of the results of the analysis depend on the applicability of these assumptions to the particular case under study. The method of ANOVA is no exception, but nothing in the way of doubt is raised by any of the characteristics of the results.

Of fundamental importance, and implicit in the use of an ANOVA model, is the assumption that the errors given by the residuals will be normally distributed with a zero mean and variance (σ^2). If this is not so, then the model is not correct in that it does not include all the significant variables. Accordingly, the analysis of residuals was carried out in some detail. The results support the validity of employing the four-way ANOVA in this study on three counts: (1) the assumption that the errors (E_{ijkm}) are normally distributed with a zero mean and a variance appears to be sound; (2) the assumption that the variances are equal among the populations of hexagons and among the populations of the locations of sample points appears quite reasonable; and (3) the ANOVA model appears to be a good postulate, that is, an analysis of the total variation among the D values, which includes the effects of the shape and size of the unit areas, the size of the hexagons, and the character of sample point locations, appears to be a sound approach.

This study does not presume to investigate the total error characteristics of the isopleth mapping process, which would be a most difficult task; instead, it was designed to analyze the effects of four of the major significant mapping variables. Nevertheless, the smallness of the mean squares of the errors (residuals) as shown in the ANOVA results (Tables 5 to 8) does suggest that the ANOVA model includes all the variables of much importance. Future research, however, should be directed toward additional exploration of these variables, and especially the large portion of the variation now assigned to the sample point location, in order to determine more completely to what extent and in what ways the operation of this factor affects the accuracy of isopleth mapping.

Magnitudes of Map Discrepancies

The subsequent analyses of the experimental data focus primarily on the resultant maps as units, instead of on the D values at each sample point, which are dealt with in the ANOVA. The magnitude of the error of an isopleth map is hereafter expressed by the average of the absolute values of D, and this average is then compared with similarly derived averages for the other maps. This is done at four different levels: (1) between the map derived from any one original surface and those derived from the other three, (2) among the four various unit area patterns associated with each surface, (3) among the five hexagonal patterns associated with each surface, and (4) among various combinations of surfaces, unit area patterns, and hexagon sizes.

The average discrepancies of the maps prepared on

the basis of various unit area patterns and hexagon sizes are shown in Figure 17. The increase of error magnitudes in the maps derived in order from Surfaces I to IV, like the progression of the complexity of these surfaces, is vividly shown. The maps based on Surface IV have by far the largest average discrepancies. On the other hand, the errors of the maps derived from Surfaces I and II are generally small, with the former being slightly better.

One may question the direct comparison of the absolute magnitudes of the errors among the four original surfaces, since a particular amount of deviation may be very critical to a certain locality but not to another locality on the same surface (or on a different surface). For example, in the case of the land surface, a given amount of hypsometric error may be considered as being more important in an area of small local relief than in an area of large local relief. Nevertheless, the magnitudes of these errors are expressed by averages rather than by individual D values, and since the comparison is made on general rather than specific terms it is not improper.

In all four cases the largest average error among the maps of various unit area patterns within a given surface is produced by Pattern IV, which consists of unit areas of variable shape and size (Figure 11). But the average discrepancies associated with the various patterns, shown by the magnitudes of the changes in the slopes of the curves on Figure 17A, differ considerably from surface to surface.

It should be reiterated that the effect of the variations induced by the chance sample point locations is included in the data shown on these graphs, and some portion of the differences of the slopes of the curves is probably due to this factor.

The magnitude of the average error generally increases with the increase in the size of hexagon, that is, with the decrease in the number of control points (Figure 17B). There is a notable exception, however, in that the variation of the hexagon size does not seem to affect

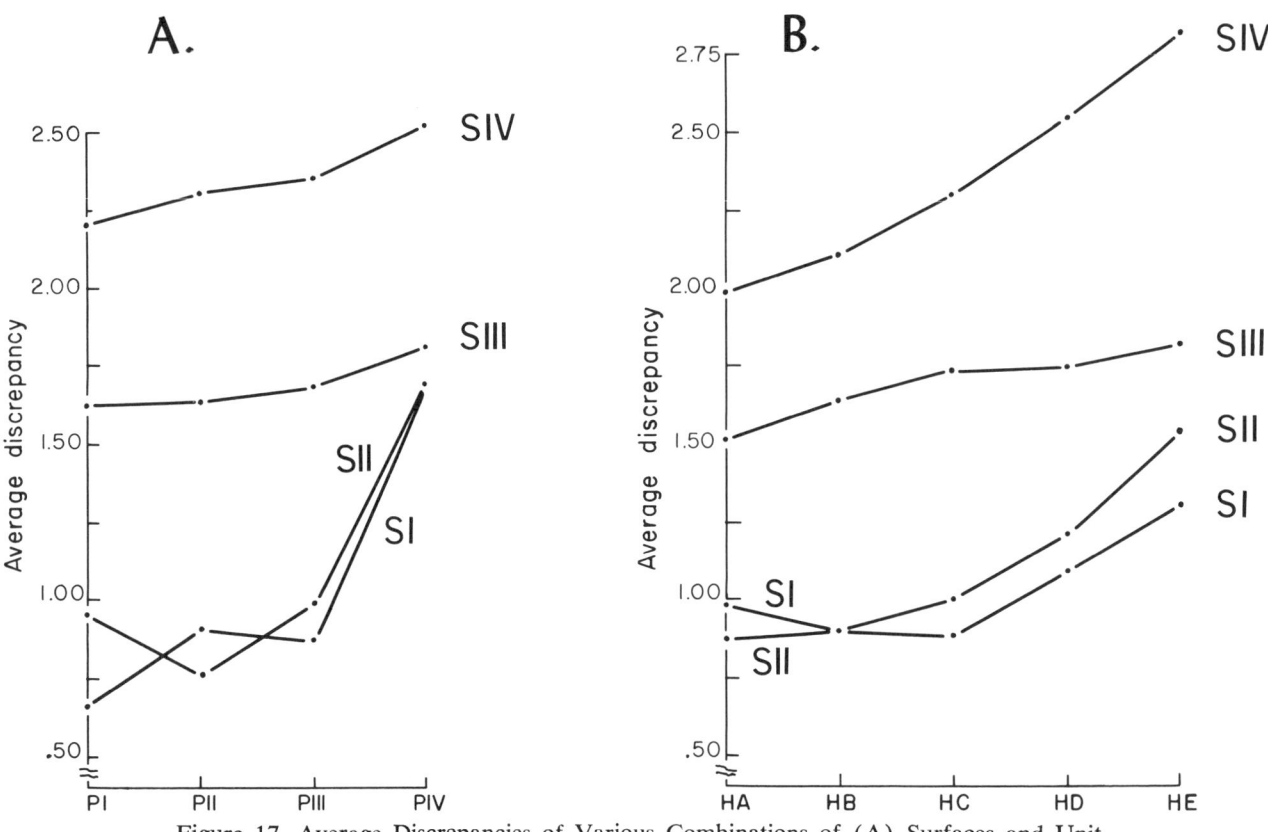

Figure 17. Average Discrepancies of Various Combinations of (A) Surfaces and Unit Area Patterns and (B) Surfaces and Hexagonal Patterns

the magnitudes of the errors of the maps derived from Surface III in the same way that it affects those derived from the other surfaces. No explanation is offered.

Finally, the twenty individual isopleth maps derived from each surface are compared with one another. In order to show graphically the interrelations among the average discrepancies and various combinations of surfaces, unit area patterns, and hexagonal patterns, the averages of each of these four sets of maps are plotted on the vertical scales of four isometric graphs shown in Figure 18. Again it is apparent that error magnitudes increase from Surface I to Surface IV, as shown by the contrast among the heights of the four graphs.

With respect to the maps derived from Surface I, Figure 18 shows that the average errors of the twelve maps based upon combinations of PI, PII, and PIII, and HA, HB, HC, and HD are all relatively small; of these, SI-PI-HA, SI-PI-HB, and SI-PI-HC yield the smallest errors among all eighty resultant maps. These three maps are derived from the simplest surface, a uniform unit area pattern, and small to medium sizes of hexagons. (Hexagon C is equal to the size of the average unit area of the patterns; HA is one-fourth of HC and HB one-half of HC.) Discrepancies increase rapidly as one moves out in both x and y directions from these twelve maps. HC seems to be the optimum size of hexagon, in the sense that it is a more expedient choice than either HA or HB, yet it produces comparable map accuracy.

The graph comparing the average errors of the isopleth maps derived from Surface II is similar to that for Surface I with three minor exceptions: (1) the magnitudes of the discrepancies are generally higher for Surface II, (2) the error consistently increases with increasing size of hexagon, and (3) Pattern II, rather than Pattern I, yields the smallest discrepancies. The first two exceptions are predictable and understandable, but the third one cannot now be explained.

The graph of Surface III, like the curves of Surface III on Figure 17, varies the least within the classifications of unit area and hexagonal patterns. The variable effect of the sample point location is probably one of the important causes of this phenomenon, for the correlation coefficient between E and D values (Table 3) is the highest for this surface. Although the configuration of the graph is rather flat, its height does indicate that, overall, the discrepancies of maps derived from Surface III are higher than those of maps derived from Surfaces I and II.

The graph of the average errors of maps derived from Surface IV is an interesting one. Here the differences in the errors associated with the various unit area patterns are relatively small compared to the differences associated with the various sizes of hexagons. This probably is related to the configuration of Surface IV, which has complex spatial variations but no distinctive distributional trends. A large number of control points would be needed to preserve the numerous details, and therefore the number of control points plays a critical role with respect to the level of map accuracy. Consequently, in terms of producing more precise maps, the smaller hexagons are far better than the larger ones. When very large hexagons are used, the details disappear or are "filtered out," and they cannot, of course, be regained on the resultant isopleth maps.

Dispersion of Discrepancies

In the preceding section the magnitudes of the errors on the resultant maps were reviewed by comparing the average discrepancies among the various classifications. In this section the dispersions of the errors are reviewed by comparing the standard deviations from the averages.

The averages were calculated from the absolute values of the errors rather than from their algebraic values because, as previously observed, in this instance the absolute mean is a more realistic measure. By ignoring the signs, one does, however, obtain a statistic that shows less variation within a group of discrepancies. Thus the standard deviation ($s_{|x|}$) about a mean error of the absolute values is either equal to or, more often, smaller than the standard deviation (s_x) about the algebraic mean for the same map.

For purposes of comparison, both types of standard deviations are calculated. The same isometric graph method is used to portray the differences among the standard deviations (Figures 19 and 20). On these two sets of graphs, the difference between the values of $s_{|x|}$ and s_x of the same series of maps is shown by the contrast in the heights of the corresponding graphs. Quite apart from this contrast, however, one can see that the forms of the corresponding graphs strikingly resemble each other. Thus both types of standard deviations increase with the complexity of the surface, the variations of the patterns of the unit areas, and the sizes of the hexagons.

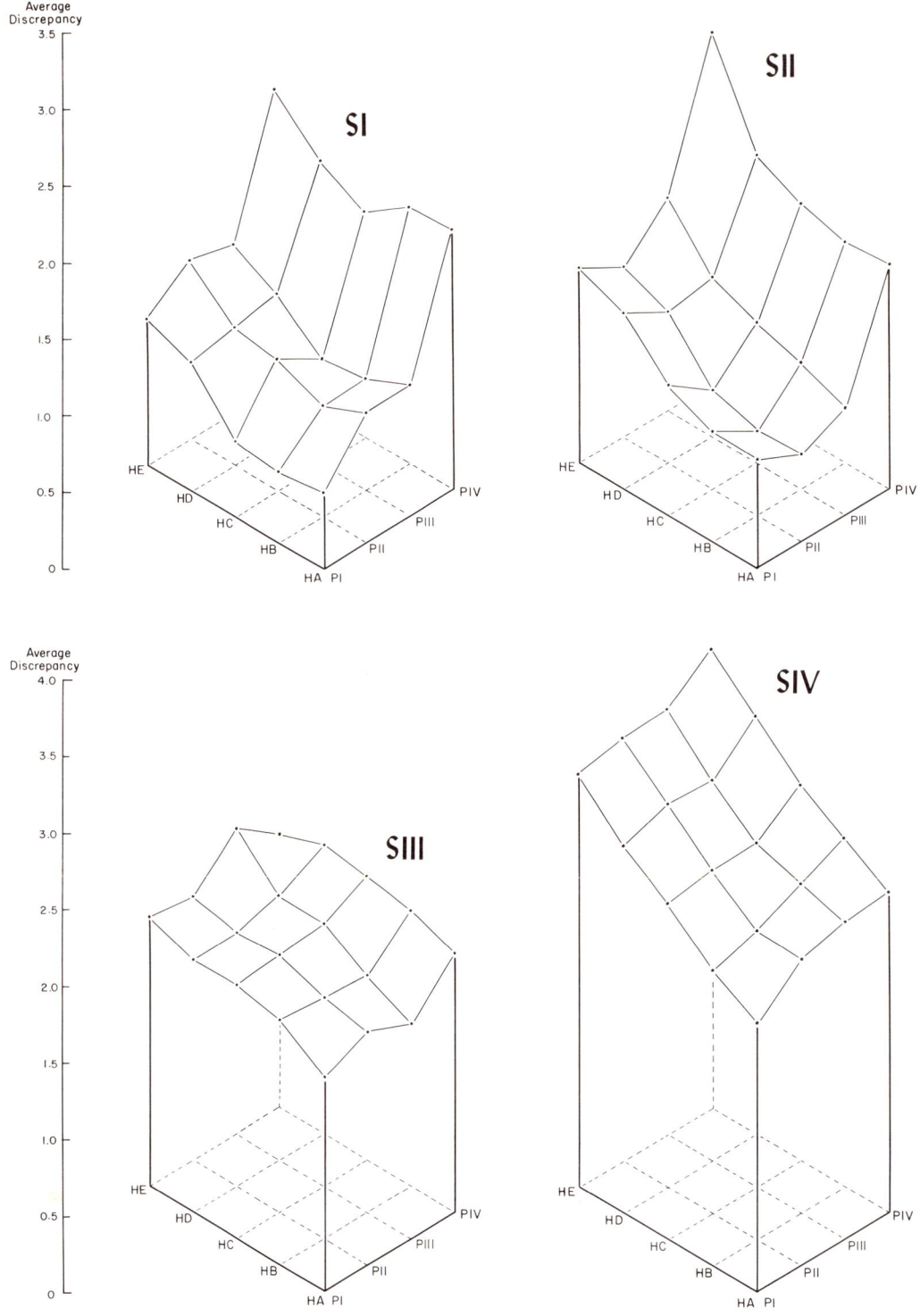

Figure 18. Interrelations among Average Discrepancies of Isopleth Maps Arranged by Surfaces, Unit Area Patterns, and Hexagon Sizes

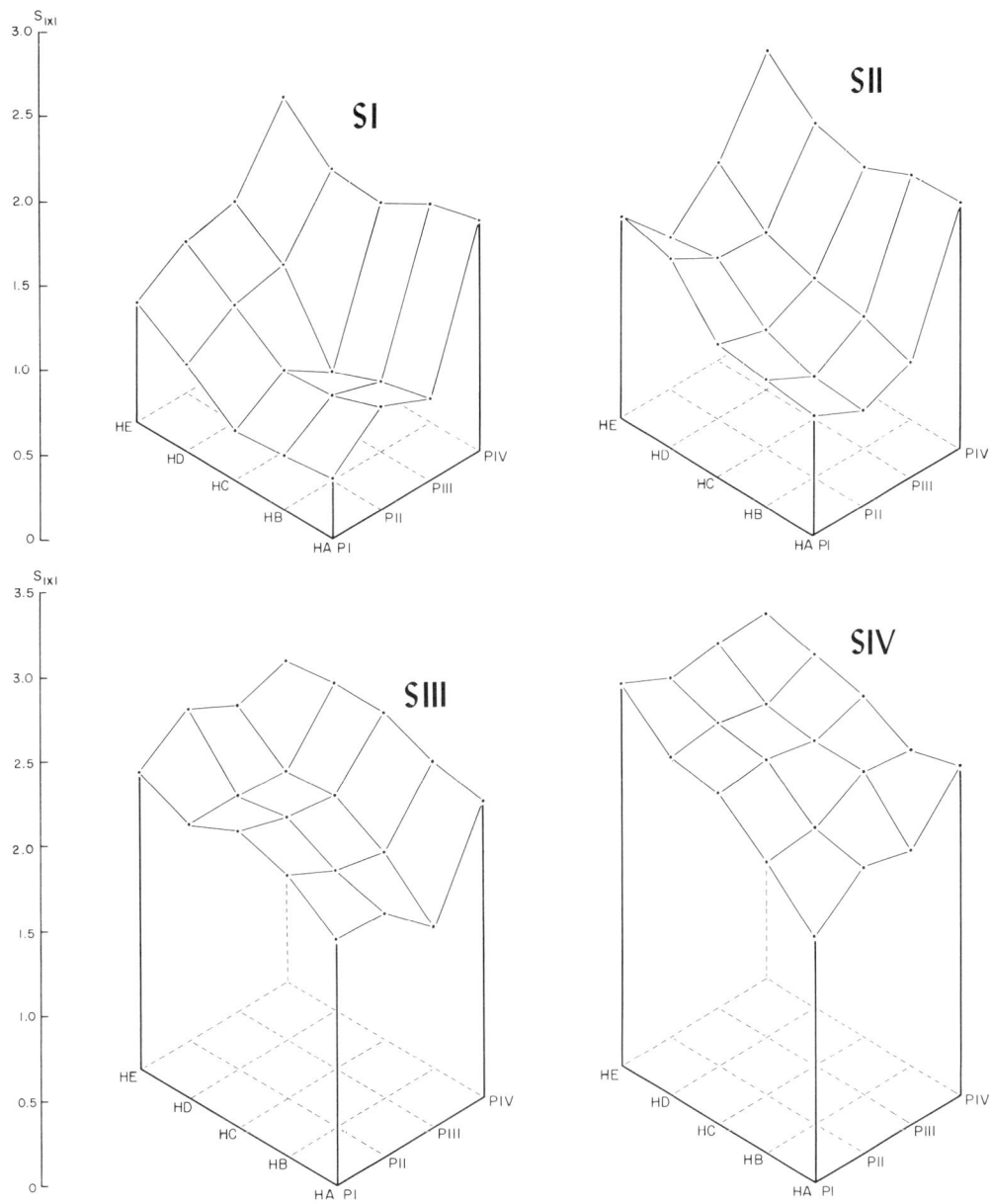

Figure 19. Standard Deviations about Averages of Absolute Values of Discrepancies on Isopleth Maps, Grouped by Surfaces, Unit Area Patterns, and Hexagon Sizes

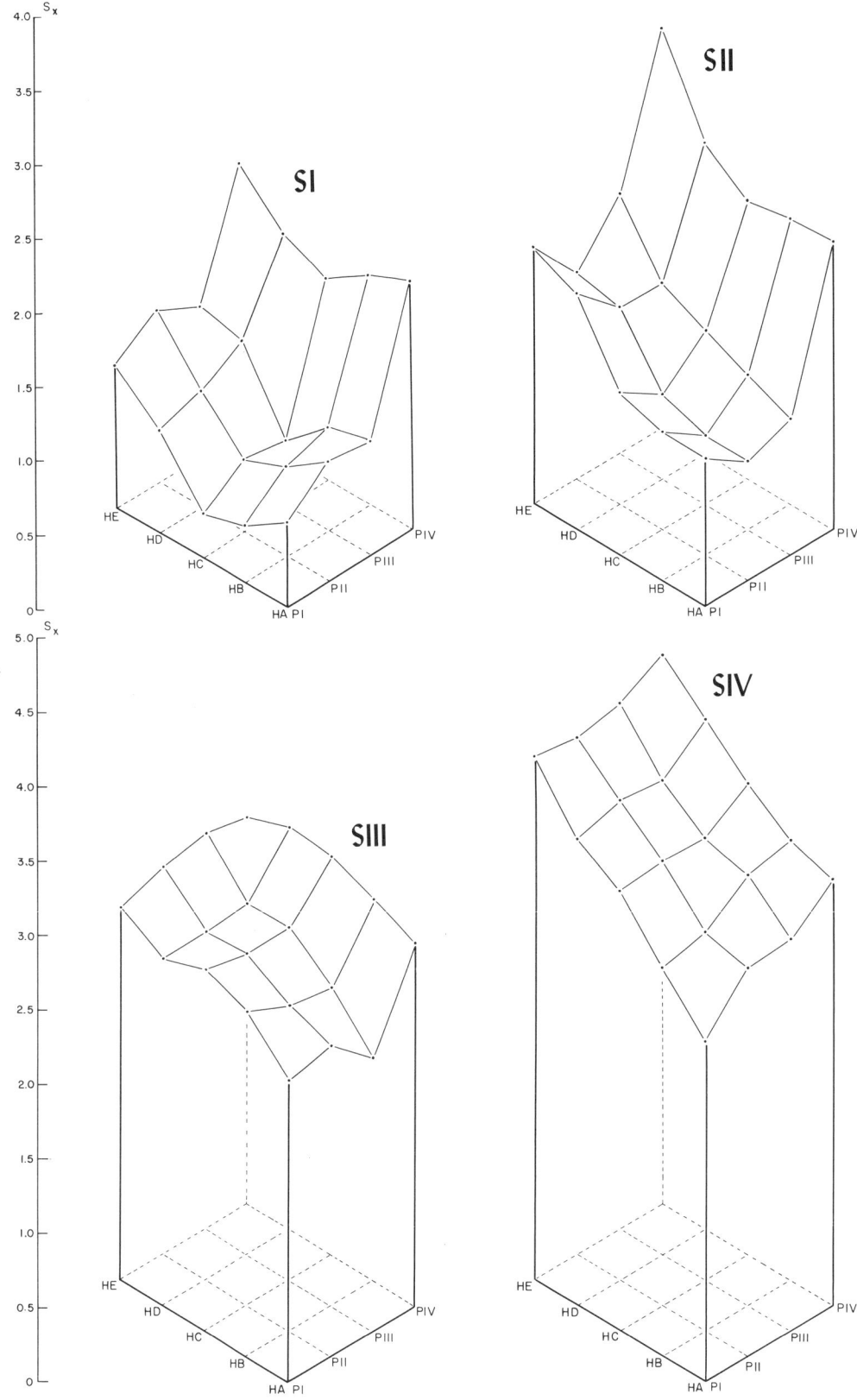

Figure 20. Standard Deviations about Averages of Algebraic Values of Discrepancies on Isopleth Maps, Grouped by Surfaces, Unit Area Patterns, and Hexagon Sizes

STATISTICAL ANALYSES

Figures 18 and 19 together show that the resultant isopleth maps which have large average discrepancies are also likely to have large standard deviations about the means. In order properly to compare the degrees of dispersion, therefore, it is helpful to employ the relative measure, the coefficient of variation, that is,

$$\text{c.v. }(\%) = \frac{s_x}{\bar{x}} \times 100.$$

The coefficients are summarized in Table 9. It is apparent that there is no significant relation between the varying magnitudes of the coefficients of variation on the one

Table 9. Average Coefficients of Variation of Errors of Isopleth Maps

Isopleth Map	Average c.v. (Percentage)
Surface	
I	69.0
II	94.0
III	98.8
IV	82.9
Unit Area Pattern	
I	87.3
II	88.5
III	82.2
IV	86.8
Hexagon	
A	87.0
B	88.8
C	84.8
D	85.4
E	85.1

hand and the categories of unit area patterns and hexagon sizes on the other; for example, the average coefficients of the four unit area patterns are quite close. Nevertheless, the average coefficients do vary a great deal among the maps derived from the various surfaces. The coefficients of Surfaces I, II, and III vary directly with the progression of surface complexity, but that of Surface IV does not fit this progression for some as yet unknown reason.

Algebraic Values of Discrepancies

Thus far the analysis of the errors of the resultant maps has generally ignored their algebraic signs; these signs or the types of discrepancies are worth noting, however. The terms "fidelity" and "map accuracy," as used in this study, are defined as the ability of an isopleth map to represent the z values of a given volume distribution, and this ability is measured here by noting the differences between the original values and the map values at sample points, the D values. A positive difference indicates that the resultant map overrepresents the z value at a certain point, and a negative difference indicates underrepresentation of the original z value.

In order to ascertain whether there is any determining factor(s) in, or any systematic arrangement of, the occurrence of the two types of errors (that is, positive or negative), a count was made of the number of positive discrepancies in the total of 30 samples on each isopleth map, and these are summarized by original surfaces in Table 10. (The maximum possible average occurrence is 30. There are a very few zero discrepancies, which were ignored in the count in Table 10.)

Table 10. Average Occurrence of Positive Discrepancies on Each Map

Surface	Average Number
I	4.2
II	12.5
III	18.9
IV	16.2

In most cases, as might have been expected, the division between positive and negative errors is not exactly 50/50. Except for the maps derived from Surface I, the ratio varies only moderately from map to map within the group derived from one surface; for example, the number of positive errors ranges from 15 to 21 among all the isopleth maps based on Surface III, and from 15 to 18 for those based on Surface IV. For maps of Surface I, however, there is an unusually small number of positive discrepancies.

Although the average occurrence of positive discrepancies shown in Table 10 varies from surface to surface, it is obvious that three of the four figures are fairly close to 15, which would be the average occurrence if the proportion between the positive and negative discrepancies had been exactly 50/50 or very close to it. In order to verify this, it may be hypothetically assumed that the division of the two types of errors is 50/50 and that the observed departures from that ratio are due only to sampling error. This assumption was then checked by a chi-square test. The results indicate that the probability is at least 0.90 that the ratio of the types of errors is 50/50 for the maps derived from Surfaces II, III, and IV, but the hypothesis must be rejected with respect to the maps derived from Surface I. In that case, the probability is less than 0.01 that the proportion is actually 50/50. Further discussion of this point will appear later.

As was pointed out earlier in Chapter 3, the variation among the magnitudes of the discrepancies at one particular sample point on all the twenty isopleth maps derived from one surface is often smaller than the variation among the errors of the thirty samples on a single map. It appears not only that the discrepancies at a given sample point have relatively consistent magnitudes, but also that they tend to have consistent algebraic signs. For example, when all the sample point discrepancies are grouped in three classes according to their signs—namely, (1) 80 percent or more of the discrepancies at a sample point are positive, (2) 80 percent or more are negative, and (3) the remainder—only 15 percent of the sample points fall in the third class, and most of these have rather small values. This suggests that small errors are likely to fluctuate around the zero value.

Apparently the consistency, noted previously, of the division between the positive and negative discrepancies from map to map of the group of isopleth maps derived from the same surface results from the similarity of the signs of the errors at each sample point. Therefore, if there is some set of factors that tends to induce a given type of map error, it is likely that it functions in the same way as the phenomenon of sample point location, namely, independently of the classifications of unit area patterns and hexagon sizes. On the other hand, the proportions of positive and negative errors do differ among the surfaces (Table 10). It appears likely, then, that the characteristics of the original surface, and more specifically the particular location of a sample point on that surface, have a systematic effect on the occurrence of the types of discrepancies.

Surface I descends from its highest point at the central right margin in all directions with a constant gradient (Figure 4). The errors on the isopleth maps are overwhelmingly negative, and the chi-square test clearly rejects the hypothesis that the probable occurrence of the two types of errors is 50/50. Furthermore, the few positive D values occur mostly along the left margins of the maps. (The points in question are 1, 3, 4, 5, and 6, shown in Figure 15.) This is interesting because the only conceivable change in the surficial characteristics of Surface I is the gradual reduction of the rate of curvature of the isarithms from the right to the left margins. Whether there is in fact a general correlation between the signs of the discrepancies and the rate of surface curvature is not known, but it is obvious that in this case the isopleth maps derived from Surface I generally underrepresent the original distribution.

Surface II (Figure 5) is irregular but similar to Surface I, and in average positive error per map it ranks between Surface I and the more complex Surfaces III and IV; however, the chi-square test indicates a high probability for an even proportion of positive and negative errors for all maps derived from Surfaces II, III, and IV.

A more certain relation between the algebraic sign of an error and the character of the surface at a sample location is illustrated by Surfaces III and IV (Figures 6 and 7). On the former, there are three distinctive "ridges" with "troughs" between them. All but one of the thirty sample points have D values that are consistently positive or negative; that is, only one point has less than 80 percent of its errors either positive or negative. Without exception the points having mostly positive errors lie in the troughs, and the points having mostly negative errors are found along the ridges. Thus, if a profile transverse to the grain of the surface is drawn along a segment of the profile between a maximum and a minimum (that is, between the peak of a ridge and the bottom of a trough), the discrepancies at the sample points which fall near the maximum are likely to be negative and those at the sample points which fall near the minimum are likely to be positive.

The association between the sample point location and the types of errors that appear in the maps derived from Surface III is also found in 85 percent of the sample points on maps derived from Surface IV. Furthermore, it is significant that the maps based on Surface IV (which is complicated by a large number of "ups" and "downs" but lacks distinctive linear trends) generally shows a more balanced occurrence of over- and underrepresentation of the z values (Table 10).

Apparently, there is a close association between the sign of the error at a particular point on the resultant maps and its sample point location on the original surface. Because of the level of generalization used in this study, the isopleth maps have lost a great deal of the details, that is, the variations, of the original surfaces. Consequently, the maps are likely to overrepresent the z values of points located in or near areas of lower z

values (that is, lower than the immediate surroundings), because the lower values are averaged out in the generalizing process and thus disappear. Conversely, the isopleth maps are likely to underrepresent the z values of points in or near areas of higher values. The spatial variations of Surfaces II, III, and IV are irregular, however, and the surface character at various sample locations is variable; consequently, this may generate a nearly even occurrence of positive and negative errors. On the other hand, the systematic surface character of Surface I is quite similar at each sample point and therefore tends to lead persistently to one type of error.

The fact that the phenomena associated with the sample point location are related to the occurrence of positive and negative discrepancies can be shown more specifically. In Chapter 3 the level of generalization at a sample point was approximated by a value of C, and a value termed E was defined as $E = C - G$, where G is the original z value at the point. The correlation analysis between the absolute values of E and D showed that there is a positive association between these two types of values. Since E measures approximately how much an original value is misrepresented on the isopleth maps as a consequence of generalization, its algebraic sign should also suggest the direction of misrepresentation. Therefore, at each sample point the algebraic sign of the E value and that of the majority (80 percent or more) of the discrepancies should be the same if there is an association between the type of discrepancy and the process of generalization.

For all surfaces, and for 70 percent of the total number of sample points, the sign of the majority of the errors is the same as that of the corresponding E value. For the remaining 30 percent, half consist of sample points the errors of which have mixed signs (that is, no majorities, as defined previously). Thus there is clearly an association between the factors involved in the generalization of the areal data and the occurrence of positive and negative discrepancies.

It may be observed that although theoretically a level of generalization should, by itself, have no effect on the occurrence of types of discrepancies, in practice it does. For example, at a low level of generalization an isopleth map of a distribution would contain most of the details, although it would not exactly reproduce the original distribution. (In terms of this experiment this would lead to a group of small discrepancies with mixed signs at each sample point; that is, theoretically there would be as many positive as negative discrepancies at each point.) As the level of generalization increased, however, more and more of the local details of higher and lower values of parts of the distribution would be filtered out in the generalizing process, and thus the unique surface character around a sample point would become an increasingly critical factor in influencing the type and the magnitude of the isopleth map error at that point. At the levels of generalization employed in this study, one observes that the dominant types and consistent magnitudes of discrepancies occur over and over at the various sample points.

In summary, when the level of generalization is quite low, it should have very little effect on the occurrence of types of errors on the isopleth maps. But with an increase in the level of generalization, the localized variation of the surface becomes an important factor in determining the types of errors.

5

Visual Analysis of the Fidelity of the Isopleth Maps

ONE way of judging the quality of a map is to define the terms of "accuracy" and then measure the map with these characteristics in mind. Properly analyzed, such quantities result in summaries of statistical terms like those developed in the previous chapter. These are meaningful and useful, for they provide the basis for the development of rigorous generalizations, as well as making available quantitative measures of probability of error that may be useful in analyses made from data derived from isopleth maps. An analysis of the fidelity of isopleth maps that stopped at that point would be incomplete, however, since for some purposes straightforward quantitative measures are not sufficient.

Isopleth maps are used for many purposes, not the least important of which is simply to provide a representation of the general form and gradients of a distribution. Many important generalizations in geographical studies have been based on distributions characterized by isopleth maps in which the total integrated, qualitative elements of the general form — that is, a sort of perceptual summarization of the vector characteristics — are of prime importance. This aspect of fidelity is more easily judged visually than by measurement. A program to this end is carried out simply by comparing the resultant isopleth maps with the original distributions from which they are derived. The results of the comparisons are arranged in three groups: first, by groupings of the unit area patterns; second, by groupings of the hexagon sizes; and last, by distribution complexity as characterized by the differences among the four original surfaces.

A map is said to be visually better than another map of the same surface if it appears to resemble the original distribution more closely, that is, appears to retain more of the basic characteristics of the original distribution. The judging was done by the authors rather than by a group of respondents in a test situation, like the one employed by McCarty and Salisbury.[1] The latter approach is probably more objective theoretically, in the sense that it presents collectively a number of individual judgments which are then summarized as the average visual reaction. But it is too elaborate for two reasons. First, even if the comparisons are made by a large group of people, the results can of course be accepted only with reservations since a visual study, like all analytical methods, has limitations.[2] Second, in this study the problems involved in visual comparison are rather simple: all maps derived from any one original distribution are presumed to be positively correlated, and all the maps derived from each surface also have the same scale and isopleth interval. Thus in our view comparative observations can be made quite easily and accurately.

Effect of the Unit Area Pattern

The F tests of the ANOVA of all surfaces indicate unanimously that variations in both the sizes and shapes of the unit areas making up the patterns do affect the fidelity of the resultant isopleth maps. If the "significance" of the F tests is to be evident visually, then for each original surface the isopleth maps which are reproduced from the various unit area patterns should "look different"; for example, map SI-PI-HA should look different than SI-PII-HA.* Note that for a fair comparison the hexagon size should be held constant.

For the isopleth maps based on Surface I, no consistent observable difference occurs between the maps derived from Pattern I and those derived from Pattern III; that is, SI-PI-HA is similar to SI-PIII-HA, SI-PI-HB is similar to SI-PIII-HB, and so on (Plates 1 and 3). Both sets of maps retain the basic characteristics of the isarithms of Surface I, which are actually segments

---

* The reader is reminded that the designation SI-PI-HA indicates the map derived from data provided by the combination of Surface I, unit area Pattern I, and Hexagon size A. In addition, the abbreviated designation SI-PI refers to the group of five maps with the same combination of surface and unit area pattern but different hexagon sizes. Finally, SI-PI-HA, HB, and HC refers to the three maps, SI-PI-HA, SI-PI-HB, and SI-PI-HC.

of a series of concentric circles. Similar as they appear, however, the maps made by way of Pattern I are slightly better than those of Pattern III, for some isopleths on the maps of Pattern I are less irregular than their counterparts on maps of Pattern III.

The group of maps derived from Pattern II and Surface I (Plate 2) is similar to the group derived from Patterns I and III; however, there are some noticeable disagreements. First, the upper right sections of the maps of Pattern II are either left blank or filled with extrapolated isopleths (in the illustrations such isopleths are shown by short dashes). Second, in the central upper portions of the maps of Pattern II the isopleths trend straight upward rather than turning to the right as do the isarithms on Surface I and the isopleths on the maps derived via Patterns I and III. The probable explanation of these differences is found on Pattern II itself.

On Pattern II there are a few large unit areas in the upper half of the pattern, with the largest one, Unit Area 5 (abbreviated hereafter as UA5) being located in the upper right corner. Since there is only one unit area value for such a large area, when a group of small hexagons is superimposed and the data transformed, there can be only one identical hexagonal value for all hexagons of that group. Therefore on SI-PII-HA, HB, and HC the corner is blank, that is, there is no spatial variation within the defined interval in that area; this, of course, is not true of the original surface. This phenomenon is relatively less obvious on the maps of larger hexagons, however. UA4 of Pattern II is also relatively large and elongated, with the longer dimension along the y axis. Thus it artificially creates a larger gradient along the x axis between its control value and those of the adjacent unit areas and a smaller gradient along the y axis. These contrasts of gradients are then transplanted, partially if not entirely, into the hexagonal data. Since there is a larger gradient in the x direction, the isopleths are necessarily developed into a more or less vertical orientation.

The well-known fact that extraordinarily large unit areas have the effect of reducing, or averaging out, the spatial variations in a distribution suggests the converse: a group of extremely small adjacent unit areas might promote relatively more variation. Therefore isopleths derived from such data may be expected to be more wiggly and intricate. This effect is not clearly observable on the resultant maps, probably for two reasons: there is no marked concentration of a group of small unit areas on any of the four patterns, and the unit area data have been transformed to hexagonal data, which applies a kind of "uniform" level of generalization. The effect of the artificial variations introduced by small unit areas would to a certain extent be spread out among various hexagons.

Finally, the isopleth maps derived via Pattern IV from Surface I are quite distinctive in comparison with the corresponding maps derived by way of the other three unit area patterns (Plate 4). First, large blank areas are found here and there in the upper and left sections of the maps; second, the isopleths in the left halves of the maps tend to have a horizontal orientation and terminate at the left margin rather than to form parabolas open to the right; and third, the isopleths in the central upper section of the maps also take a vertical orientation (this is most obvious in SI-PIV-HA).

The probable reasons for these features are again found in the pattern itself (Figure 11). The unit areas in the left half of Pattern IV are large and elongated, with the longer dimension along the x axis. As seemed to be the case with the maps of Pattern II, the blank areas are probably induced by the large unit areas, within which much of the surface variation is filtered out; and the horizontal orientation of the isopleths is likely due to the presence of the elongated unit areas, which create artificial gradient contrasts, and lead the isopleths to such an orientation. For the same reason UA2 in the central upper portion of Pattern IV is probably responsible for the upright orientation of the isopleths in that area.

Since Surface I is a regular form, it is easy to observe the contrasts which seem to be induced by the use of various unit area patterns. On the other hand, it is more difficult to analyze the irregularities produced by the unit area patterns on the resultant maps of the variable Surfaces II, III, and IV. For instance, one cannot be sure whether the vertical orientation of the isopleths in the central upper portion of the maps of SII-PII (Plate 6) is affected by UA4 of Pattern II, as is evidently the case in the maps SI-PII, since the isarithms in that portion of Surface II also have a vertical orientation. Another example is found in the maps of SIII-PII and SIII-PIII (Plates 10 and 11). Both groups of maps show blank areas in their upper right sections. On the former, these are probably caused by the large unit area (UA5) on Pattern II (mentioned earlier in connection with the maps of SI-PII), but another reason must be sought for their presence on the maps of SIII-PIII. There are two medium-size unit areas in the upper right of Pattern III (Figure 10), and it so happens that their aggregated unit area values derived from Surface III (that is, in the combination SIII-PIII) are identical; thus, according to the unit area data, there is no spatial variation in that area. This obviously has been carried over into the hexagonal data and reflected on the resultant maps.

If one looks beyond the compound irregularities that result from the variable surfaces and unit area patterns, he will find that the observations made earlier about Surface I seem to apply to all the resultant maps, regardless of the surface of origin. In summary, then, the effects of the unit area patterns on the isopleth maps are as follows.

1. The maps made via Patterns I and III are visually similar, and the former are somewhat better than the latter. More generally, the resultant isopleth maps may be ranked in the order of Patterns I, III, II, and IV in terms of their resemblance to their respective original surfaces.

2. On all maps made by way of Pattern II, the upper right section is either left blank or filled with extrapolated isopleths. This is particularly evident on the maps of SII-PII. Here the highest values of the surface are located in that section and they are averaged out in deriving the aggregated value for the large UA5. Consequently, the upper right portions of maps of SII-PII are not only left blank but generally have lower maximum values in comparison with their corresponding maps derived via other patterns, such as SII-PI, SII-PIII, and so forth.

3. The maps made via Pattern IV are the poorest. On the left halves of the maps, the blank areas mentioned in connection with maps derived from SI-PIV become even larger and more conspicuous in the series SII-PIV, SIII-PIV, and SIV-PIV. This is particularly obvious in the series SIV-PIV (Plate 16), where the upper left sections, about one-fifth of the entire area of the map, are blank. On these maps that area is portrayed as an even distribution. Needless to say, this is not the character of the original surface (Figure 7). Furthermore, the peculiar but consistent horizontal orientation of the isopleths on the left halves of the maps appears prominently on all the maps derived via Pattern IV. This is a good example of the way that the variations of an original distribution may be completely obscured by the effects of the size, shape, and arrangement of the unit areas used in isopleth mapping.

Effect of the Hexagon Size

The F tests of the ANOVA of all surfaces showed that different hexagon sizes induce different levels of map accuracy. Visual comparisons were also carried out on the maps derived from various hexagon sizes. (For this comparison the unit area patterns were, of course, held constant.) In all cases the maps of various hexagon sizes do indeed look different. The gradual loss of detail from the maps based on the smallest hexagons to those based on the largest is evident. When a resultant map has lost a great deal of the original surface variation, it apparently can no longer retain the major distributional characteristics of the surface. For example, one can hardly trace the linear concentrations on some of the resultant maps prepared from Surface III.

The probable "level" at which the loss of detail becomes critical to the visual quality of a map cannot, unfortunately, be easily ascertained. Naturally, this depends in part on the use to which the map is being put, and it is affected not only by the differences of hexagon sizes but also by the contrasts among unit area patterns and surfaces. Generally speaking, a critical level does not appear to occur at all in the series derived from Surfaces I and II. For the maps developed from Surfaces III and IV, the level appears occasionally in maps made via Hexagon C, but it is always present on maps made via Hexagons D and E. Any variation appears to be due largely to the differences of the qualities of the unit area patterns.

These observations should not lead one to infer that the maps of the smallest hexagons are necessarily the best. In fact a peculiar phenomenon is presented on maps of SI-HA. On SI-PI-HA (Plate 1), for example, there are four groups of isopleths across the central part of the map, each containing three closely adjacent lines. Together they depict a steplike surface which, of course, does not exist on Surface I. The cause becomes apparent when one examines the data from which the map was prepared. Hexagon A is one-fourth the size of the average unit area, that is, one-fourth of a unit area of Pattern I (which has unit areas of uniform shape and average size). When the hexagonal pattern was ran-

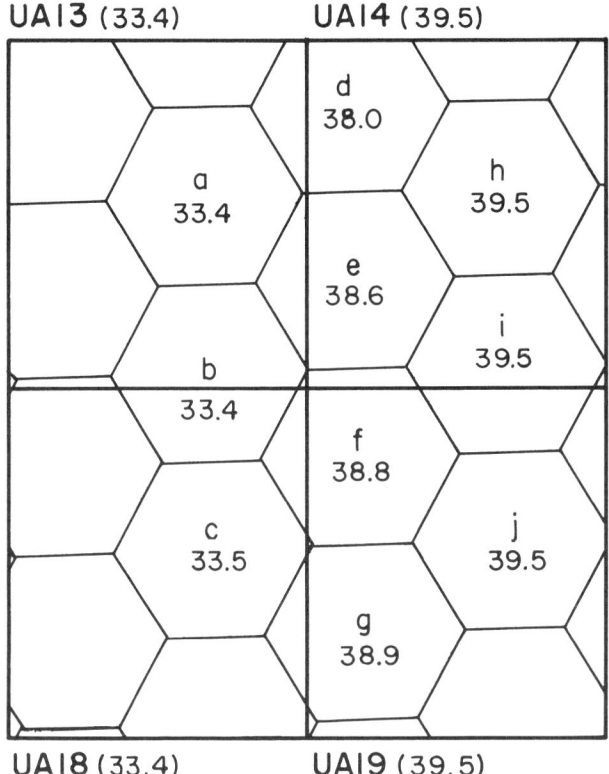

Figure 21. Section of Four Unit Areas from Superimposition of HA on SI-PI

domly superimposed on Pattern I, by chance two columns of hexagons covered approximately one column of unit areas (Figure 21). Therefore, the differences between the hexagon values of a-d and a-e are greater than those between d-h and e-h. The contrasts in gradient are further reinforced by the fact that on SI-PI the value of UA13 happens to equal that of UA18, and the value of UA14 equals that of UA19, whereas the values of the two pairs are quite different because of the arrangement of the isarithms of Surface I (see Plate 1). In circumstances such as these, it is likely that the isopleths between a-d and a-e would be greater in number and more closely located than those between d-h and e-h. In fact, on SI-PI-HA no isopleth was placed between d-h and e-h but three were drawn between a-d and a-e. The arrangement shown in Figure 21 is not an unusual occurrence.

When the entire series of isopleth maps based on Hexagon A is judged, there is a greater difference between those derived on the one hand from Surfaces I and II, and on the other from Surfaces III and IV. In the first group, the maps of HA are naturally more detailed than those made via the larger hexagons, but more often than not the detail is falsely created by the smallness of the hexagons themselves. The random bunching of the isopleths just described is a good example. The isopleths of the maps of SI-HA are, in general, too wiggly in comparison with the original surface; and although the maps of SII-HA look impressively detailed, the basic trend of many of the isopleths is at variance with the nature of that surface. There seems no question but that maps based upon the largest number of control points are not the best ones for Surfaces I and II. On the contrary, among the maps of Surface I, those based on HC seem to have the highest visual quality; and for Surface II the maps based on HB seem to be the best. These maps are more generalized than those of HA but do contain a reasonable amount of information to indicate the basic characteristics of the original surfaces.

The second group of maps based on Hexagon A tells another story. Surfaces III and IV are quite complex, and, at the level of generalization used in this study and regardless of the hexagon size, one must conclude that no resultant isopleth maps are good representations of these two surfaces. Certainly, the maps based on the smallest hexagons are the best by far in preserving the surface variations. Theoretically, the false wiggles and artificial contrasts that were discussed in connection with the first group of maps of HA should also be present on the maps of SIII-HA and SIV-HA. In reality, however, they are not visible at all; and as a matter of fact the isopleths are all too smooth in comparison with the isarithms on the original surfaces. In short, among the resultant isopleth maps made from Surfaces III and IV, it is quite clear that the larger the hexagon size (or the smaller the number of control points) the poorer is the resultant map.

Effect of the Surface Configuration

In the statistical analysis a separate ANOVA was run for each original surface, and therefore there are no quantitative measures by which one may easily judge whether the differences in the complexity of the surface configuration in any way affected the quality of the resultant isopleth maps. The effect of this factor on the fidelity of the isopleth process is, however, clearly revealed by a visual examination of the resultant maps.

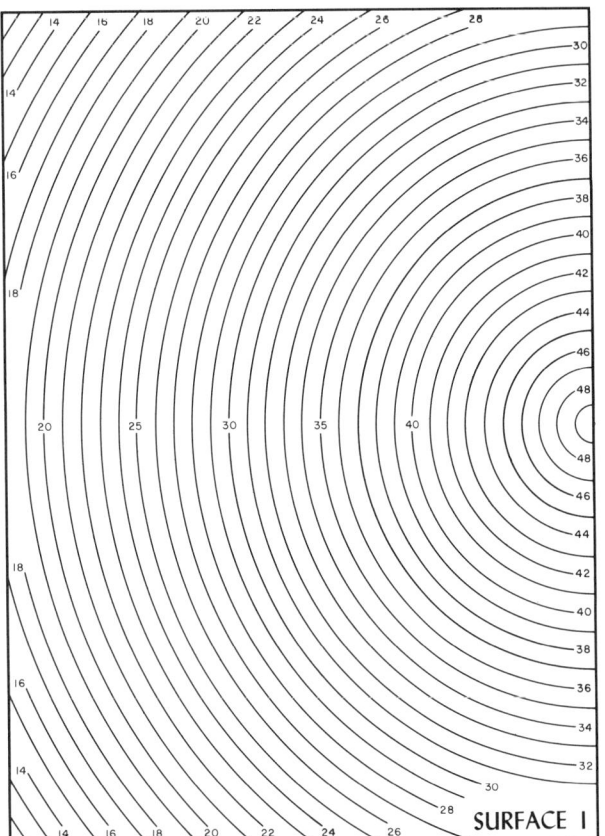 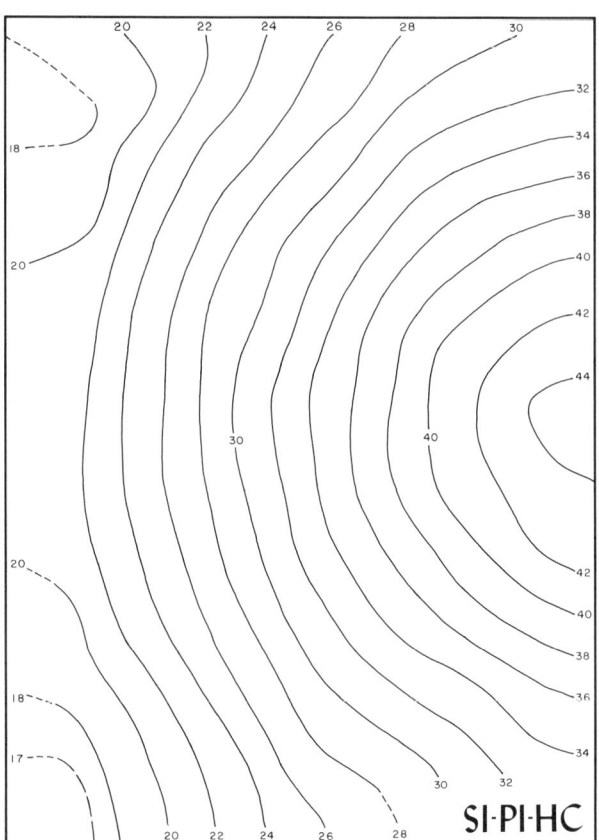

Figure 22. Surface I and Its "Best" Isopleth Representation, SI-PI-HC

For a given level of generalization, it is abundantly clear that only certain amounts of the surface variations and basic features of a distribution can be preserved on isopleth maps.

In terms of the complex of spatial character judged visually, the following isopleth maps seem to provide the best reproductions of each of the original surfaces: SI-PI-HC, SII-PI-HB, SIII-PI-HA and SIV-PI-HA (Figures 22 to 25). These maps probably represent the best that can be done, quantitatively and qualitatively, within the restrictions imposed by the prescribed mapping procedure. A comparison of these maps shows clearly that the most complex distribution suffers the greatest loss. Thus, at a given level of generalization and with a given mapping procedure, the fidelity of an isopleth map derived from a complex surface configuration is likely to be very much lower than that of one derived from a simple surface.

As for Surface I (the simplest distribution), with the exception of maps of SI-PIV all the isopleth maps retain its basic characteristics — namely, a unimodal surface with the peak at the central right margin. The only distinctive but minor departure from the original surface is found at the upper and lower left corners of the maps, where a small number of isopleths tend to curve to the left rather than to the right as do the rest of the isopleths. This is presumably due to the fact that there are fewer control points at the marginal areas and the locations of the isopleths are less certain. Comparatively, the five maps of SI-PIV represent the original distribution rather poorly. The locations of the isopleths on the left halves of the maps are adversely affected by the unit area pattern, as was discussed previously; thus they misrepresent the surface. The remaining parts of the maps, however, retain the basic features of Surface I.

Similar comments can be made for maps based upon Surface II with, probably, one reservation: the basic characteristics of the distribution tend to fade away on maps employing the largest hexagon size, SII-HE. The fact that the isopleth maps made from SI and SII have comparable qualities, of course, comes as no surprise, since the surfaces are similar.

The quality of the maps based upon Surfaces III and IV is considerably poorer. First, the great loss of original detail is very obvious, and it occurs even on the maps prepared by way of the smaller hexagons. Second, for these surfaces the visual contrast among the maps based on different hexagon sizes—that is, on various numbers of control points—is more distinctive than for Surfaces I and II. While the maps based upon the smaller hexagons do suggest vaguely the characteristics of the original distributions, the reproductions made by way of the larger hexagons bear almost no resemblance to their original surfaces (Plates 9 to 16).

Relative Significance of the Variables

It is evident from visual analysis that the characteristics of the surfaces, the sizes of the hexagons, and the characteristics of the unit area patterns are all variables which strongly affect the fidelity of isopleth maps. If one considers a volume distribution in terms of its surface complexity and distributional trends, as has been done in this study, the effects of the first two variables—the characteristics of the original configuration and the size of the hexagons employed in the transformation—are the most critical with respect to the retention of the amount of the spatial variations and their arrangements on the resultant maps. As pointed out previously, the isopleth maps derived from a simpler distribution lose, relatively, a smaller amount of the original spatial variation than do those derived from a more complex surface. Similarly, the resultant maps made by way of smaller hexagons preserve much more of the detail of the original surface than do those based on the larger hexagons. The conclusions about the effect of various sizes of hexagons may, of course, be generalized to the sizes

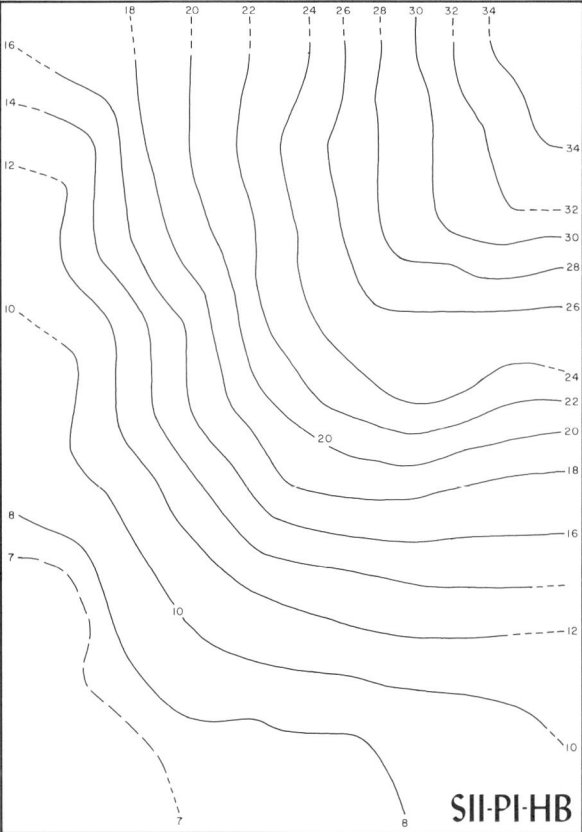

Figure 23. Surface II and Its "Best" Isopleth Representation, SII-PI-HB

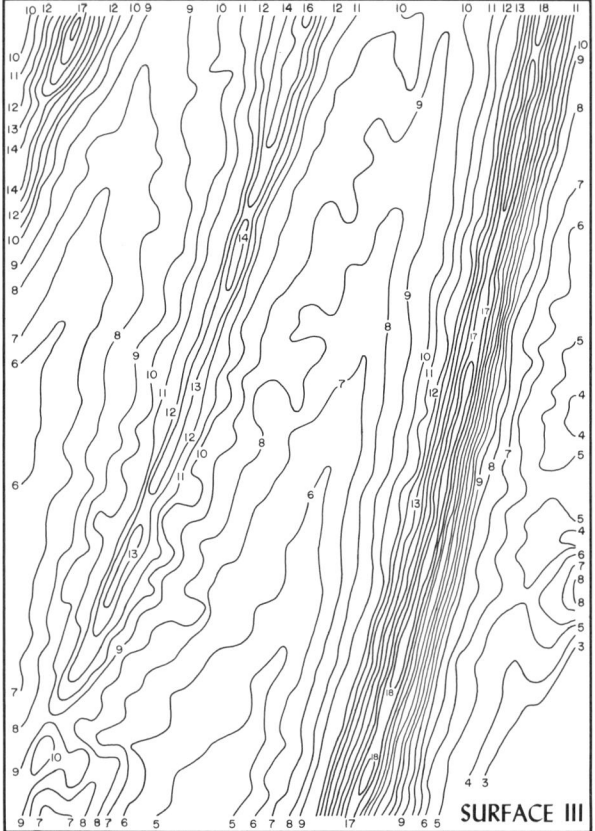
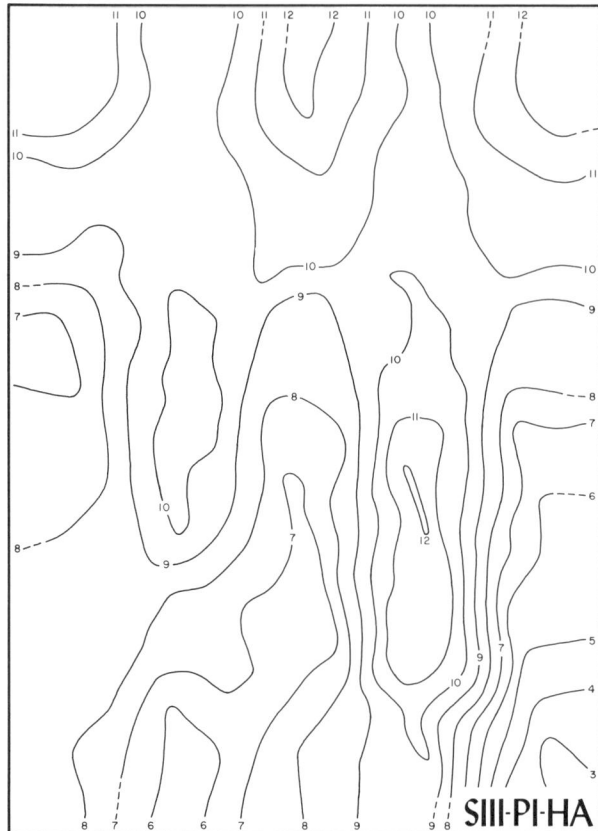

Figure 24. Surface III and Its "Best" Isopleth Representation, SIII-PI-HA

of the cells in any other transformation patterns, for example, rhombuses, rhomboids, or rectangles.

The third variable — the characteristics of the unit area patterns — appears to be the most critical with respect to the retention of the "vectoral quality" of the distributional variations and arrangements on the original surfaces, that is, the magnitudes and directions of gradients. Consequently, this variable affects the appearance of the resultant maps quite differently than does the complexity of a surface configuration or the size of the hexagons. As was observed earlier, the variable sizes, shapes, and arrangements of the unit areas can easily produce certain sets of data which radically alter the gradients and directions of the isopleths on the resultant maps. Consequently, isopleth maps can definitely misrepresent the general form of an original surface. Since the general form of the configuration presented on a small-scale isopleth map is often more important than the intricate details of the isopleths and the precise values at places, one may conclude that the characteristics of the patterns of unit areas employed are very critical to the fidelity of the isopleth process.

While a variation in the size of the unit areas seems to affect mainly the representation of the magnitudes of the surface gradients on the resultant maps, a variation in shape apparently can actually change the direction of the gradient.[3] As is illustrated by maps made via Pattern IV, where a group of large elongated units is present, the isopleth maps so derived show a false distribution; see, for example, the resultant maps of Pattern IV (SI-PIV, SII-PIV, and so forth). The critical element is apparently not the minor irregularity of the shape of a unit area but its departure from the most optimum form, a circle. This is evident from the fact that maps developed from Pattern III (Figure 10) are relatively good.

In establishing the experimental design, it was argued that it was advantageous to transform the unit area data to hexagonal data and to use the latter as the basic

Figure 25. Surface IV and Its "Best" Isopleth Representation, SIV-PI-HA

data for drawing the isopleth maps. It was pointed out, among other things, that this intermediate step reduces the effect of the variation of the pattern of unit areas on the quality of the resultant maps. Presumably this reduction has occurred, although not to the extent expected. In some cases the influences of the shapes and sizes of unit areas are so powerful that their imprint remains strong even after the transformation of the data from the variable unit areas to uniform hexagons. This certainly raises the question of the usefulness of such a transformation (or any similar processes employing other geometric shapes), if the main purpose of the transformation is to reduce significantly the effect on the isopleth map of the variations in the size, shape, and arrangement of the enumeration districts employed to aggregate the data.[4]

A more thorough way of minimizing the effect on the isopleth mapping process of a variable pattern of unit areas would be to disaggregate the unit area data and re-enumerate them by employing a more ideal pattern, that is, a uniform pattern of unit areas. The critical task then would be the process of disaggregating the unit area data. This inevitably leads one to suggest a fundamental change in the existing method of census-taking by adopting a geometric pattern for the enumeration districts (such as a hexagonal pattern of relatively small units or divisions defined by latitudes and longitudes), instead of using the existing administrative units. A computer program could be employed to transform the census data enumerated upon such a geometric pattern to administrative units, and therefore two sets of "censuses" would be available for different uses.

6

General Conclusions

AS WE have shown, the fidelity with which an isopleth map portrays the character of the real distribution from which it is derived is affected by (1) the character of the distribution being mapped, (2) the sizes, and (3) the shapes of the unit areas employed to enumerate the data, and (4) the number of control points upon which the map is based. In this study the number of control points was a function of the nature of the transformation, but with or without transformation one must always work with a number of control points. This is established either by the number of unit areas or by the character of the transformation, and in any case it is closely related to the degree of generalization.

Differences between the isopleth maps and the original surfaces at sample positions provided discrepancy or error values which collectively may be considered to represent the relative accuracy of the isopleth maps. The analysis of variance procedure to which the experimental data were subjected was run once for each "detailed distribution," that is, for each of the four original surfaces from which isopleth maps were derived. The ANOVA results show that all the main effects are statistically significant in controlling isopleth accuracy. When the relative importance of the main effects are compared among the four original surfaces, the information provided by the ANOVA indicates that, of the other three variables, the size of the unit areas is the most critical factor influencing the variations of the accuracy of the isopleth maps derived from Surface I. Shape of the unit areas is the most critical factor for maps derived from Surface II. The most critical factor for the maps derived from Surfaces III and IV turned out to be the chance location of the sample points. For all surfaces except Surface I, shapes of the unit areas are more influential than their sizes.

The error terms in the ANOVA, the residuals, are all very small, indicating that for each surface the variables under consideration did indeed account for the overwhelming portion of the variations in the degree of accuracy among the resultant isopleth maps.

When the average discrepancy of each isopleth map is calculated and the averages grouped according to various combinations of surfaces and unit area patterns, it is evident that for each surface (with minor exceptions) the averages are the smallest on maps derived from unit area Pattern I and the largest on those from unit area Pattern IV. Similarly, within the group of maps derived from each surface, the average errors of the maps prepared via the smallest hexagon size are generally the smallest, and those from the largest hexagon size, the largest. When comparisons are made among the four surfaces, the isopleth maps derived from the simplest surface, SI, have the smallest average errors and the errors increase in order from SI to SIV. When the effect of various surfaces, unit area patterns, and hexagon sizes are considered collectively, the smallest discrepancies are usually found on those maps made by way of combinations of simple surfaces, unit areas with uniform size and shape, and medium to small hexagons.

The occurrence of positive and negative discrepancies at the thirty sample points on each isopleth map is reasonably close to a chance division for maps derived from Surfaces II, III, and IV but not for those derived from Surface I. The variations of unit area patterns and hexagon sizes do not seem to affect the occurrence of the positive or negative type of map error. But there is evidence suggesting that, at least at the level of generalization employed in the study, the configuration of the original surface and the location of the sample point do affect the occurrence of this type of map error. Along a profile transverse to the grain of the original distribution, a sample point which falls near a "ridge" is likely to have a negative error while one near a "trough" is likely to have a positive error.

Visual evaluation of isopleth fidelity confirms most of the results of the numerical analysis, but with some exceptions. For example, as judged visually, the maps based on the four different unit area patterns rank in the order of PI, PIII, PII, and PIV for all surfaces; but the general order would be PI, PII, PIII, and PIV on the quantitative basis of average discrepancies.[1]

Both the analysis of group means in the ANOVA and

GENERAL CONCLUSIONS

the comparisons among the average discrepancies of maps derived by way of the various hexagon sizes indicate that maps produced via smaller hexagons generally have smaller errors. The visual quality of the isopleth maps also declines with an increase in the size of hexagon. The increasing loss of the distributional details from maps made with the smallest hexagon to those made from the largest is quite apparent, and this is more obvious for maps derived from Surfaces III and IV than for I and II. In addition, the visual comparison of the maps derived from different hexagon sizes contributes at least two important points supplementary to the numerical analysis. First, while maps derived from Surface III are quite similar in the magnitudes of their average discrepancies regardless of hexagon size, those derived by way of Hexagons A and B are far better in appearance than those of D and E. Second, while the maps derived from Hexagon A usually have the smallest average discrepancies, they do not necessarily possess the best visual quality.

With the mapping procedure employed in this experiment, the more complex original distributions appear to lose relatively a larger amount of spatial variations, and therefore the resultant maps have a comparatively lower level of accuracy. The maps derived from Surface I retain the basic characteristics of that surface (except for SI-PIV), as do the maps produced from Surface II (except that maps of SII-HE seem to begin to lose the essential character of that surface). Clearly, however, the maps based on Surfaces III and IV generally appear poor. While the ones derived from the smaller hexagons and from unit area patterns with less variable cells vaguely suggest the characteristics of the original surfaces, those reproduced from less favorable unit area patterns and hexagonal patterns in no way resemble the original distributions.

Analysis of the differences in appearance leads to the conclusion that the characteristics of the original distribution (surface) and the number of control points (sizes of hexagons) affect mainly the quantitative aspects of the spatial variations which can or cannot be preserved by the isopleth mapping process. On the other hand, the kinds of patterns of unit areas are more critical in affecting preservation of the surficial character of the spatial variations, such as ridges and troughs and directions of gradients. In fact, in some instances the impact of the variations in the shapes and sizes of unit areas is so strong that their effects are carried through from the unit area data to the hexagonal data. This raises a question as to the usefulness of such cartographic transformation.

If one attempts to choose from the eighty isopleth maps the four which seem to characterize the original distributions with the greatest fidelity, it becomes clear that no single variable is overpowering. The authors' selection of SI-PI-HC, SII-PI-HB, SIII-PI-HA, and SIV-PI-HA (Figures 22 to 25) as the best that the isopleth process can produce indicates that, while the simpler distributions "come through" rather well, the more complex ones lose a great deal. Certainly the isopleth maps representing Surfaces III and IV are poor substitutes for reality, and a geographer could rely on only their grossest aspects.

At this stage of our knowledge, it is safe to conclude that the quality of isopleth mapping is greatly affected by the complexity of the original distribution, the variation in the pattern of unit areas (sizes and shapes), and the number of control points used. It is still not possible, however, to define precisely the manner in which these variables affect the fidelity of a map. For example, one cannot prepare a comprehensive summary which includes a complete classification of surfaces, unit area characteristics, and hexagonal patterns on the one hand and the probable measures of the quality of the isopleth maps on the other. The major problem seems to be that the essential spatial characteristics of quantitative distributions and unit area patterns have not yet been numerically defined, and further understanding is needed concerning the effects of sample point location and generalization of areal data on the precision of the isopleth map. In all instances, but particularly with respect to Surfaces III and IV, the effect of the sample point location clearly turned out to be significant in explaining the variations in degrees of accuracy among the resultant maps. This is a very important technical problem in the research design, for it is fundamentally a source of error (sampling error?) in obtaining the experimental data. If instead of using a sampling method in order to obtain point values of map errors, the accuracy of a resultant map were evaluated by comparing the volume beneath the original surface with that of an isopleth representation, the problems due to sample point locations could be eliminated. On the other hand, the kinds of data available for analyses would be greatly reduced.

Table 11. Points Assigned to Various Surfaces, Unit Area Patterns, and Hexagon Sizes

Category	Points Assigned	Category	Points Assigned	Category	Points Assigned
Surface		Pattern		Hexagon	
I	1	I	1	A	1
II	2	II	2	B	2
III	4	III	2	C	3
IV	5	IV	4	D	4
				E	5

Further, this research has not explored specifically the relation between the level of generalization and the quality of a resultant map. It would certainly be useful to know the optimum level of generalization for the purpose of the best description for each "type of surface," that is, quantitative geographical distribution.

A small attempt is made here, however, to show graphically the multiple relation between these variables and the fidelity of isopleth maps. On the basis of the relative findings of this research, an arbitrary point system has been developed in which the determinants having greater complexity and variation are assigned higher numbers (see Table 11). Each surface, unit area pattern, and hexagon size is then assigned an arbitrary value, and a point total is allocated to each resultant map. For example, map SI-PI-HA received a total of three points and SI-PIV-HD nine points. Then the point total for each map is paired with the average quantitative error of each map and these are plotted on a scatter diagram (Figure 26). To be sure, such a system is arbitrary; nevertheless, the diagram does show a trend of modest positive correlation between the various combinations of surfaces and patterns on the one hand and map errors on the other. At the present stage the use

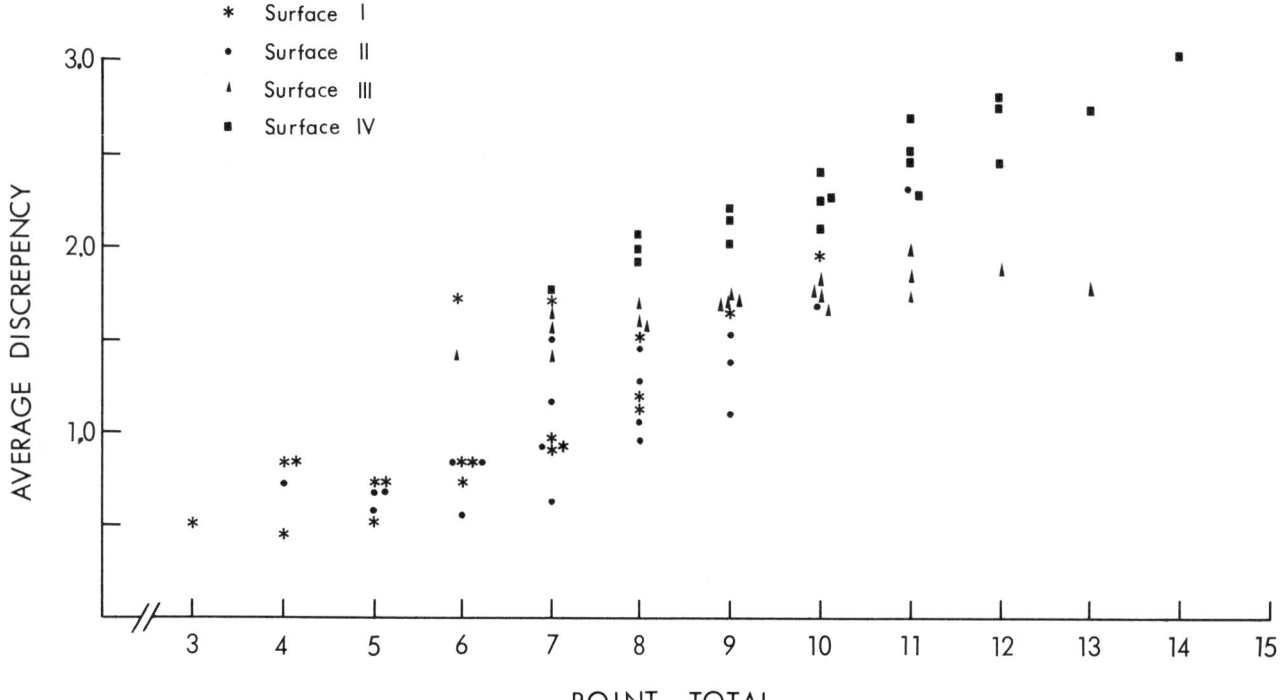

Figure 26. Scaling Diagram of Average Discrepancies of the Isopleth Maps Made from the Four Surfaces Plotted against an Arbitrary Summation of Points Based on Surface, Unit Area Pattern, and Hexagon Size

GENERAL CONCLUSIONS

of more elaborate statistical methods such as multiple correlation analysis is not likely to yield additional information.

Perhaps it is fitting to conclude this analysis by observing that the complexity of the isopleth mapping process appears to be great, and our basic knowledge of it limited, and therefore only certain tangible relationships can now be observed. In that light Figure 26 is revealing, both in terms of the information it conveys about the eighty resultant maps and as a graphic statement calling for further research and understanding of the cartographic process.

Notes

Notes

Chapter 1. Quantitative Mapping and the Isarithm

1. S. J. Fockema Andreae and B. van 't Hoff, *Geschiedenis der Kartografie van Nederland* (The Hague: Martinus Nijhoff, 1947), p. 95 and plate 15.
2. H. F. van Riel, "Pierre Ancelin," *Tijdschrift voor Kadaster en Landmeetkunde*, 40 (1925), 51–56, 133–44; François de Dainville, "De la profondeur à l'altitude," *International Yearbook of Cartography*, 2 (1962), 151–60.
3. See Dainville's article, cited above, for a full treatment of this development in France, where the contour had its first widespread, officially sanctioned use.
4. G. Hellmann, "Magnetische Kartographie in historische-kritischer Darstellung," *Veröff. d. königlich Preuss. Met. Inst.*, no. 215, Abh. Bd. III, no. 3, pp. 5–61 (Berlin, 1909).
5. S. Chapman, "Edmund Halley as a Physical Geographer and the Story of His Charts," *Occasional Notes*, Royal Astronomical Society, 9 (1941), 122–34.
6. A. H. Robinson and Helen M. Wallis, "Humboldt's Map of Isothermal Lines: A Milestone in Thematic Cartography," *Cartographic Journal*, 4 (1967), 119–23.
7. For the first isoline maps of population density, see Niels F. Ravn, "Populations Kaart over Det Danske Monarki 1845, Ditto, 1855," in *Statistik Tabelvaerk*, n.s., vol. 12 (Kon. Dan. Statist. Bureau, Copenhagen, 1857); for isoline maps of agricultural production, see A. Petermann and E. Behm, "Die Verbreitung der hauptsächlichsten Kultur-Produkte in den Vereinigten Staaten von Nord-Amerika," *Petermanns Geographische Mitteilungen*, 2 (1856), 408–39, plates 20–35.
8. For a general review of the history of these kinds of lines in mapping, see W. Horn, "Die Geschichte der Isarithmenkarten," *Petermanns Geographische Mitteilungen*, 103 (1959), 225–32; for a listing of the various ways the lines have been used, see J. L. M. Gulley and K. A. Sinnhuber, "Isokartographie, eine terminologische Studie," *Kartographische Nachrichten*, 11 (1961), 89–99.
9. Charles Dupin, *Forces productives et commerciales de la France*, 2 vols. (Paris: Bachelier, 1827), vol. 2, plate 1.
10. The first carefully done dasymetric map seems to have been made by Henry Drury Harness in 1837. See Arthur H. Robinson, "The 1837 Maps of Henry Drury Harness," *Geographical Journal*, 121 (1955), 440–50 and George F. McCleary, Jr., "The Development of the Dasymetric and Other Similar Concepts in Thematic Cartography" (Ph.D. diss., University of Wisconsin, 1969).
11. Arthur H. Robinson, "The Cartographic Representation of the Statistical Surface," *International Yearbook of Cartography*, 1 (1961), 53–61. For a discussion of the objectives and techniques of the three classes of statistical mapping, see Robinson and Randall D. Sale, *Elements of Cartography*, 3rd ed. (New York: John Wiley, 1969), pp. 142–46.
12. See W. D. Jones, "Isopleth as a Generic Term," *Geographical Review*, 20 (1930), 341; John K. Wright, "The Terminology of Certain Map Symbols," *Geographical Review*, 34 (1944), 653–54; Robinson and Sale, *Elements of Cartography*, pp. 156–57.
13. For a discussion of the classification of geographical quantities, see John K. Wright, "Crossbreeding Geographical Quantities," *Geographical Review*, 45 (1955), 52–65. For an account of the application of isopleth mapping, see Philip W. Porter, "Putting the Isopleth in Its Place," *Proceedings of the Minnesota Academy of Science*, 25–26 (1957–58), 372–84.
14. The technique of plastic shading need not be limited to portraying the land surface. See Robinson, "Cartographic Representation of the Statistical Surface," pp. 59–61.
15. There are those who argue that a ratio involving area ought not be represented cartographically by a contour-like symbol. The reason put forward is that since, for example, population density — number of persons per unit of area — is a ratio involving area, it cannot occur at a point. An isopleth "joins all points of equal value" and consequently suggests an impossibility. It seems to the authors that this argument is empty and stands in a class with Zeno's paradox that one cannot ever reach a point because he must always first go halfway. It is fundamentally a matter of the degree to which one will think abstractly. Certainly, the wide use of the isopleth in the more than a century that has elapsed since its first application is evidence enough of its utility.
16. The term "error characteristic" implies that the errors so introduced are generally random, in terms of both their occurrence on the map and their magnitudes; that is, the amount of error is likely to vary randomly rather than persistently. Therefore, generalization is not an error characteristic, since it can be argued that generalization should be introduced at a consistent level throughout an ordinary map.
17. J. Ross Mackay, "An Analysis of Isopleth and Choropleth Class Intervals," *Economic Geography*, 31 (1955), 71–81; *idem*, "Isopleth Class Intervals, a Consideration in Their Selection," *Canadian Geographer*, 7 (1963), 42–44; George F. Jenks and M. R. C. Coulson, "Class Intervals for Statistical Maps," *International Yearbook of Cartography*, 3 (1963), 119–34; Robinson and Sale, *Elements of Cartography*, pp. 164–70.
18. W. Z. Ripley, "Notes on Map Making and Graphic Representation," *Quarterly Publication of the American Statistical Association*, 6 (1898), 310–26; John K. Wright, "Map Makers are Human, Comments on the Subjective in Maps," *Geographical Review*, 32 (1942), 527–44; J. Ross Mackay, "Some Problems and Techniques in Isopleth Mapping," *Economic Geography*, 27 (1951), 1–9; George F. Jenks, "Generalization in Statistical Mapping," *Annals of the Association of American Geographers*, 53 (1963), 15–26.
19. For a more detailed discussion of error characteristics of isopleth maps, see Mei-Ling Hsu, "An Analysis of Isarithmic Accuracy in Relation to Certain Variables in the Mapping Process" (Ph.D. diss., University of Wisconsin, 1966), chap. 3; Joel L. Morrison, "The Effects of Sampling and Interpolation in Isarithmic Mapping" (Ph.D. diss., University of Wisconsin, 1967).
20. David I. Blumenstock, "The Reliability Factor in the Drawing of Isarithms," *Annals of the Association of American Geographers*, 43 (1953), 289–304.
21. Fr. Uhorczak, "La Méthode Isarythmique Appliquée Aux Cartes Statistiques," *Polski Przeglad Kartograficzny*, 8 (1930), 96; Wright, "Map Makers Are Human," pp. 535–38.
22. The statement is open to argument on two grounds. First, one can indeed consider the arbitrary pattern designed by a cartographer as a sampling system which he uses to collect areal data, and he could very well design another sampling system if he desired. On the other hand, if he uses administrative units, their boundaries cannot be altered, at least not by

the cartographer. Therefore, he must work with a given set of census data, or a sampling statistic, and there is no chance of obtaining any other sample. Second, while areal data are collective values within the statistical units, nevertheless all observations of the population are enumerated. The process is not really "sampling." It may be better to think of a certain set of areal data as one of the possible populations within the larger universe, rather than as a sample of a population.

23. Blumenstock, "Reliability Factor in the Drawing of Isarithms," pp. 289–304.

24. Mackay, "Some Problems and Techniques," pp. 4–5. Mackay has illustrated how the control points can be located in areas of regular and irregular shapes.

25. E. E. Svaitlovsky and W. C. Eells, "The Centrographical Method and Regional Analysis," *Geographical Review*, 27 (1937), 240.

26. Mackay, "Some Problems and Techniques," pp. 4–5.

27. Bengt-Erik Bengtsson and S. Nordbeck, "Construction of Isariths and Isarithmic Maps by Computers," *BIT*, 4 (1964), 87–105; M. O. Dayhoff, "A Contour-Map Program for X-Ray Crystallography," *Communications of the Association for Computing Machinery*, 6 (1963), 620–22; Morrison, "Effects of Sampling and Interpolation."

28. See Morrison, "Effects of Sampling and Interpolation," chap. 3, for a discussion of the problems of interpolation models in computer isarithmic mapping.

29. Uhorczak, "La Méthode Isarythmique," pp. 100–101; Mackay, "Some Problems and Techniques," p. 5; idem, "The Alternative Choice in Isopleth Interpolation," *Professional Geographer*, 5, no. 4 (1953), 2–4.

30. Porter, "Putting the Isopleth in Its Place," pp. 375–76.

31. The term "control point density" refers to the relation between the number of control points and the size of the area to be mapped.

32. Mackay, "Some Problems and Techniques," p. 7; Robinson and Sale, *Elements of Cartography*, pp. 158–60.

33. John W. Alexander and G. A. Zahorchak, "Population-Density Maps of the United States: Techniques and Patterns," *Geographical Review*, 33 (1943), 457–66.

34. Uhorczak, "La Méthode Isarythmique," pp. 95–130.

35. Waldo R. Tobler, "A Polynomial Representation of Michigan Population," *Michigan Academy of Science, Arts, and Letters*, 49 (1964), 445–51. In this paper Tobler discusses the use of a polynomial for representing a geographical volume quantity. A polynomial, $P_n = (aX + bY + c)^n$, developed by the method of least squares is fitted to the distributional surface. The main purpose of his study was not isopleth mapping; however, as he pointed out: "A very simple computer program has been prepared which makes such maps by evaluating the descriptive equation some 5,000 times per page and then assigning appropriate symbols to class intervals." Thus the interpolation is done indirectly through evaluating the descriptive polynomial. The accuracy of the interpolation depends, necessarily, on the goodness with which the polynomial fits the surface. Since the polynomial cannot be expected to fit the surface exactly, there will still be errors introduced by this process of interpolation. See also Morrison, "Effects of Sampling and Interpolation."

36. Wright, "Map Makers Are Human," p. 539; see also Arthur H. Robinson, "Mapping the Correspondence of Isarithmic Maps," *Annals of the Association of American Geographers*, 52 (1962), 414–25, especially pp. 416–17.

37. Torsten Hägerstrand, *The Propagation of Innovation Waves*, Lund Studies in Geography, Series B, no. 4 (1952), pp. 4–7; Porter, "Putting the Isopleth in Its Place."

38. Porter, "Putting the Isopleth in Its Place"; Arthur H. Robinson, J. B. Lindberg, and L. W. Brinkman, "A Correlation and Regression Analysis Applied to Rural Farm Population Densities in the Great Plains," *Annals of the Association of American Geographers*, 51 (1961), 214.

39. D. R. Crone, "The Accuracy of Topographical Maps," *Empire Survey Review*, 12, no. 88 (1953), 64–70; Morris M. Thompson and Charles H. Davey, "Vertical Accuracy of Topographic Maps," *Surveying and Mapping*, 13 (1953), 40–48.

Chapter 2. Preparation of the Experimental Data

1. See, for example, Philip W. Porter, "Putting the Isopleth in Its Place," pp. 379–80; Robinson, Lindberg, and Brinkman, "Correlation and Regression Analysis," p. 214.

2. The average of the unit area sizes of the patterns is 2.93 square inches except for Pattern IV, which is 2.84 square inches. For the sake of convenience, 2.90 was adopted for the mid-size hexagon. For the same reason, the area of Hexagon A is taken to be 0.73 square inch rather than 0.725. It would be more logical to let Hexagon E be four times the area of Hexagon C instead of three times, but it would not be practical. If Hexagon E had been 11.60 square inches, a map of 8 inches by 11 inches would have had only seven or eight control points. The resultant maps would be highly generalized, and the task of allocating the gross errors to various causes would be quite difficult. This difficulty exists even with Hexagon E.

3. "Randomly" here refers to placing the hexagonal pattern over the unit area pattern casually and then turning it around a few times without looking at it.

4. An experiment was conducted to examine the variation in the number of control points that arises from the different orientations of the superimpositions. Hexagon C was placed randomly over Pattern II five times, and the number of control points obtained each time was as follows: 31, 30, 31, 30, and 30, with an average of 30.4. Then the same hexagonal pattern was twice placed very carefully over Pattern II with the objective of including the maximum number of control points. The results were 30 and 31, with an average of 30.5. Similar experiments were done with Pattern II and Hexagon E. The results were similar to those of Pattern II and Hexagon C.

5. The basic difference between these processes is that the moving average employs each statistic more than once. This does not occur in the method of transformation here described.

Chapter 3. Testing and Sampling Procedures

1. Lloyd E. Marsden, "How the National Map Accuracy Standards Were Developed," *Surveying and Mapping*, 20 (1960), 438.

2. Thompson and Davey, "Vertical Accuracy of Topographic Maps," pp. 40–41, 46–47.

3. *Ibid.*

4. *Ibid.*, pp. 40–47.

5. Neil E. Salisbury and Placido LaValle, "Scale Variations in Morphometric Analysis." Paper delivered at the West Lakes Division Meeting of the Association of American Geographers, November 1–2, 1963.

6. Uhorczak, "La Méthode Isarythmique," pp. 95–130; Marja Zdobnicka, "Metoda Izarytmiczna W Grafice Statystycznej," *Poklosie Geograficzne*, 1925, pp. 255–71.

7. The terms "map error" and "discrepancy" are used synonymously. The former is self-explanatory and therefore a better term, but it may be confused with the error term (residual) in connection with the analysis of variance which is dealt with in the following chapter. Thus the term "discrepancy" is introduced.

8. William G. Cochran, "Relative Accuracy of Systematic and Stratified Random Samples for a Certain Class of Populations," *Annals of Mathematical Statistics*, 17 (1946), 164–77; M. H. Quenouille, "Problems in Plane Sampling," *Annals of*

NOTES

Mathematical Statistics, 20 (1949), 355–75; Morrison, "Effects of Sampling and Interpolation." Additional references on sampling methods are provided in the Bibliography, below.

9. It is proper to observe here that the sample of 40 points (Figure 14F) is probably a poor representative of a sample of that size. On the average, of course, a sample of 40 should be a little better than a sample of 30, but the gain in precision is not necessarily worth the extra time required to collect the additional number of point values.

10. There is a minor exception to this statement: Pattern IV has 31 unit areas.

11. No attempt is made to consider the average shape of the filters. It could be represented by the modal shape of the unit areas of the four patterns.

12. It would be useful to make further investigations of the relation between degree of generalization and map accuracy as here defined. For example, it would be interesting to determine the relative accuracy of the resultant maps if 60 and 15 unit areas had been used instead of 30, and to derive a functional relation between the degree of "uniform" generalization and map accuracy.

Chapter 4. Statistical Analyses of the Fidelity of the Isopleth Maps

1. See K. A. Brownlee, *Statistical Theory and Methodology in Science and Engineering* (New York: John Wiley, 1960), chaps. 10, 13, and 14.

2. For a more detailed discussion of the validity of the ANOVA model and the analyses of group means and residuals, see Mei-Ling Hsu, "An Analysis of Isarithmic Accuracy," chap. V.

Chapter 5. Visual Analysis of the Fidelity of the Isopleth Maps

1. Harold H. McCarty and Neil E. Salisbury, *Visual Comparison of Isopleth Maps as a Means of Determining Correlations between Spatially Distributed Phenomena* (Iowa City: State University of Iowa, 1961).

2. See *ibid.*, the conclusions of various tests.

3. See Mei-Ling Hsu, "The Isopleth Surface in Relation to the System of Data Derivation," *International Yearbook of Cartography*, 8 (1968), 75–87.

4. In order to examine at least the visual effect of hexagonal transformation, some comparisons were made between a map drawn directly from the unit area data (for example, SI-PIV) and another one drawn from its derived hexagonal data (SI-PIV-HC). When the level of generalization is comparable — equivalent numbers of unit areas and hexagons are employed — the two maps are fairly similar. See also Hsu, *ibid.*

Chapter 6. General Conclusions

1. The average errors of the maps based on SII-PII are consistently smaller than those derived from SII-PI and thus present an exception to the generalization that the average map errors increase from PI to PIV for all surfaces. This phenomenon cannot, however, be observed on the maps themselves. Visually the SII-PI maps are far better than those of SII-PII.

Selected Bibliography

Selected Bibliography

Aeronautical Chart and Information Center, Geo-Sciences Branch. *Map Accuracy Evaluation, Part II, Evaluation of Vertical Map Information.* ACIC Reference Publication no. 2, 1963.

Alexander, John W. "An Isarithmic-Dot Population Map." *Economic Geography*, 19 (1943), 431-32.

Alexander, John W., and Zahorchak, George A. "Population-Density Maps of the United States: Techniques and Patterns." *Geographical Review*, 33 (1943), 457-66.

Anscombe, F. J. "Examination of Residuals." *Proceedings of the Fourth Berkeley Symposium on Mathematical Statistics and Probability*, 1 (1961), 1-36.

Bagrow, Leo, and Skelton, R. A. *History of Cartography.* Cambridge: Harvard University Press, 1964.

Bauer, L. A. "Halley's Earliest Equal Variation Chart." *Terrestrial Magnetism*, 1 (1896), 28-31.

Bengtsson, Bengt-Erik, and Nordbeck, Stig. "Construction of Isarithms and Isarithmic Maps by Computers." *BIT*, 4 (1964), 87-105.

Berry, Brian J. L., and Baker, Alan M. "Geographic Sampling." In *Spatial Analysis*, edited by B. J. L. Berry and D. Marble, pp. 91-100. Englewood Cliffs, N.J.: Prentice-Hall, 1968.

Blaut, J. M. "Microgeographic Sampling: A Quantitative Approach to Regional Agricultural Geography." *Economic Geography*, 35 (1959), 79-88.

Blumenstock, David I. "The Reliability Factor in the Drawing of Isarithms." *Annals of the Association of American Geographers*, 43 (1953), 289-304.

Brown, Lloyd A. *The Story of Maps.* Boston: Little Brown, 1950.

Brownlee, K. A. *Statistical Theory and Methodology in Science and Engineering.* New York: John Wiley, 1960.

Byron, William G. "Method of Mapping Population Distribution with Dots and Densitometer-Derived Isopleths." Ph.D. dissertation, Syracuse University, 1954.

———. "Use of the Recording Densitometer in Measuring Density from Dot Maps." *Surveying and Mapping*, 18 (1958), 41-48.

Chapman, S. "Edmund Halley as a Physical Geographer and the Story of His Charts." *Occasional Notes*, Royal Astronomical Society, 9 (1941), 122-34.

Clarke, John I. "Statistical Map Reading." *Geography*, 44, no. 204, part 2 (1959), 96-104.

Cochran, William G. "Relative Accuracy of Systematic and Stratified Random Samples for a Certain Class of Populations." *Annals of Mathematical Statistics*, 17 (1946), 164-77.

———. *Sampling Techniques.* 2nd ed. New York: John Wiley, 1963.

Creamer, Marvin C. "Isolines in Population Density Mapping." *Professional Geographer*, 10 (1958), 14-15.

Crone, D. R. "The Accuracy of Topographical Maps." *Empire Survey Review*, 12, no. 88 (1953), 64-70.

Crone, G. R. *Maps and Their Makers.* 2nd ed. London: Hutchinson's University Library, 1962.

Czekalski, Józef. "Mapa Izarytmiczna, A Obrazrzeczywisty (Próba Analizy Metody)." *Wiadomości Służby Geograficznej*, no. 3 (1933), 202-34.

———. "Kartogram a mapa izarythmiczna." *Wiadomości Służby Geograficznej*, no. 4 (1934), 467-92; French résumé, pp. 492-94.

———. "Le rôle de la méthode isarythmique dans les recherches géographiques." *Czasopismo geograficzne*, 12, nos. 3-4 (1934), 209-20; French résumé, pp. 220-22.

Dainville, François de. "De la profondeur à l'altitude." *International Yearbook of Cartography*, 2 (1962), 151-62.

Das, A. C. "Two Dimensional Systematic Sampling and the Associated Stratified and Random Sampling." *Sankhya*, 10, parts 1 and 2 (1950), 95-108.

Das Gupta, Sivaprasad. "Methods of Isopleth Mapping." *Geographical Review of India*, 19 (1957), 10-14.

———. "Some Measures of Generalization on Thematic Maps." *Geographical Review of India*, 26 (1964), 73-78.

Dayhoff, M. O. "A Contour-Map Program for X-Ray Crystallography." *Communications of the Association for Computing Machinery*, 6 (1963), 620-22.

Dupin, Charles. *Forces productives et commerciales de la France.* 2 vols. Paris: Bachelier, 1827. Vol. 2, plate 1.

Eckert, Max. *Die Kartenwissenschaft.* 2 vols. Berlin and Leipzig: Walter de Gruyter, 1921, 1925.

Fockema Andreae, S. J., and 't Hoff, B. van. *Geschiedenis der Kartografie van Nederland.* The Hague: Martinus Nijhoff, 1947.

Funkhouser, H. G. "Historical Development of the Graphical Representation of Statistical Data." *Osiris*, 3 (1937), 269-404.

Ghosh, B. "Topographic Variation in Statistical Fields." *Calcutta Statistical Association Bulletin*, 2 (1949), 11-28.

Gibson, James J. *Proposals for a Theory of Pictorial Perception.* Human Factors Operations Research Libraries Memo Report no. 13, Air Research and Development Command, USA, Bolling Air Force Base. Washington, D.C., 1953.

Gierhart, John W. "Evaluation of Methods of Area Measurement." *Surveying and Mapping*, 14 (1954), 460-65.

Glusic, Andrew M. *The Positional Accuracy of Maps.* Technical Report Number 15 (Project no. MO-011, March 1961), Army Map Service. Washington, D.C., 1961.

Greenwalt, Clyde R., and Shultz, Melvin E. *Principles of Error Theory and Cartographic Applications.* Aeronautical Chart and Information Center Technical Report no. 96, 1962.

Gullev, J. L. M., and Sinnhuber, K. A. "Isokartographie, eine terminologische Studie." *Kartographische Nachrichten*, 11 (1961), 89-99.

Hägerstrand, Torsten. *The Propagation of Innovation Waves.* Lund Studies in Geography, Series B, no. 4, 1952.

Hellmann, G. *Neudrucke von Schriften und Karten über Meteorologie und Erdmagnetismus.* No. 4. Berlin, 1895.

———. "Magnetische Kartographie in historische-kritischer Darstellung." *Veröff. d. königlich Preuss. Met. Inst.*, no. 215, Abh. Bd. III, no. 3, pp. 5-61. Berlin, 1909.

Horn, Werner. "Die Geschichte der Isarithmenkarten." *Petermanns Geographische Mitteilungen*, 103 (1959), 225-32.

Hsu, Mei-Ling. "An Analysis of Isarithmic Accuracy in Relation to Certain Variables in the Mapping Process." Ph.D. dissertation, University of Wisconsin, 1966.

———. "The Isopleth Surface in Relation to the System of Data Derivation." *International Yearbook of Cartography*, 8 (1968), 75-87.

Imhof, Eduard. "Isolinienkarten." *International Yearbook of Cartography*, 1 (1961), 64–98.

Jenks, George F. "Generalization in Statistical Mapping." *Annals of the Association of American Geographers*, 53 (1963), 15–26.

Jenks, George F., and Coulson, M. R. C. "Class Intervals for Statistical Maps." *International Yearbook of Cartography*, 3 (1963), 119–34.

Jones, Wellington D. "Ratios and Isopleth Maps in Regional Investigation of Agricultural Land Occupance." *Annals of the Association of American Geographers*, 20 (1930), 177–95.

———. "Isopleth as a Generic Term." *Geographical Review*, 20 (1930), 341.

Krumbein, W. C. *Statistical Significance of Beach Sampling Methods*. U.S. Department of Army, Office of the Chief of Engineers, Beach Erosion Board, Technical Memorandum No. 50. Washington, D.C., 1954.

———. "Experimental Design in the Earth Sciences." *Transactions, American Geophysical Union*, 36 (1955), 1–11.

Künzel, Willy. "Die Methode der räumlichen Gruppenbildung am Beispiel einer neuen Volksdichtekarte vom Freistaat Sachsen." *Mitt. Vereins der Geographen an der Univ. Leipzig*, 10–11 (1932), 37–42.

Lägnert, Falke. *The Electorate in the Country Districts of Scania, 1911–1948*. Lund Studies in Geography, Series B, no. 5, 1952.

Learmonth, A. T. A., and Pal, M. N. "A Method of Plotting Two Variables (Such as Mean Incidence and Variability from Year to Year) on the Same Map, Using Isopleths." *Erdkunde*, 13 (1959), 145–50.

Lenz, Werner. "Iso-Linien in der Geographie." *Geographische Rundschau*, 11 (1959), 323.

McCarty, Harold H., and Lindberg, James B. *A Preface to Economic Geography*, pp. 23–40. Englewood Cliffs, N.J.: Prentice-Hall, 1966.

McCarty, Harold H., and Salisbury, Neil E. *Visual Comparison of Isopleth Maps as a Means of Determining Correlations between Spatially Distributed Phenomena*. Iowa City: State University of Iowa, 1961.

Mackay, J. Ross. "Some Problems and Techniques in Isopleth Mapping." *Economic Geography*, 27 (1951), 1–9.

———. "The Alternative Choice in Isopleth Interpolation." *Professional Geographer*, 5, no. 4 (1953), 2–4.

———. "Percentage Isopleth Maps." *Professional Geographer*, 7, no. 6 (1955), 10–12.

———. "An Analysis of Isopleth and Choropleth Class Intervals." *Economic Geography*, 31 (1955), 71–81.

———. "Isopleth Class Intervals, a Consideration in Their Selection." *Canadian Geographer*, 7 (1963), 42–44.

Madow, Lillian H. "Systematic Sampling and its Relation to Other Sampling Designs." *Journal of American Statistical Association*, 41 (1946), 204–17.

Madow, W. G. "On the Theory of Systematic Sampling, II." *Annals of Mathematical Statistics*, 20 (1949), 333–54.

Madow, W. G., and Madow, L. H. "On the Theory of Systematic Sampling, I." *Annals of Mathematical Statistics*, 15 (1944), 1–24.

Marsden, Lloyd E. "How the National Map Accuracy Standards Were Developed." *Surveying and Mapping*, 20 (1960), 427–39.

Matérn, Bertil. "Forest Surveys and the Statistical Theory of Sampling — Some Recent Developments." *Fifth World Forestry Congress Proceedings* (1960), 276–81.

———. "Spatial Variation." *Medd. fr. Statens Skogsforsknings Institut* (Stockholm), 49 (1960), 1–144.

Morrison, Joel Lynn. "The Effects of Sampling and Interpolation in Isarithmic Mapping." Ph.D. dissertation, University of Wisconsin, 1967.

Mowrer, Ernest R. "The Isometric Map as a Technique of Social Research." *American Journal of Sociology*, 44 (1938), 86–96.

Neft, David S. *Statistical Analysis for Areal Distributions*. Monograph Series No. 2, Regional Science Research Institute, Philadelphia, 1967.

Pannekoek, A. J. "Generalization of Coastlines and Contours." *International Yearbook of Cartography*, 2 (1962), 53–74.

Petermann, A., and Behm, E. "Die Verbreitung der hauptsächlichsten Kultur-Produkte in den Vereinigten Staaten von Nord-Amerika." *Petermanns Geographische Mitteilungen*, 2 (1856), 408–39, plates 20–35.

Pietkiewicz, Stanislaw. "Analyse de l'exactitude de quelques cartes du XVII, XVIII et XIX siècle, couvrant les territoires de l'ancienne Pologne." *Przeglad Geograficzny*, 32 (1960), Supplement, 21–27.

Porter, Philip W. "Putting the Isopleth in Its Place." *Proceedings of the Minnesota Academy of Science*, 25–26 (1957–58), 372–84.

———. "What Is the Point of Minimum Aggregate Travel?" *Annals of the Association of American Geographers*, 53 (1963), 224–32.

Proudfoot, Malcolm J. *Measurement of Geographic Area*. U.S. Bureau of the Census, Sixteenth Census of the U.S.: 1940. Washington, D.C.: Government Printing Office, 1947.

Quenouille, M. H. "Problems in Plane Sampling." *Annals of Mathematical Statistics*, 20 (1949), 355–75.

Raisz, Erwin. *General Cartography*. 2nd ed. New York: McGraw-Hill, 1948.

———. *Principles of Cartography*. New York: McGraw-Hill, 1962.

Ravn, Niels F. "Populations Kaart over Det Danske Monarki 1845, Ditto, 1855." *Statistik Tabelvaerk*, n.s. vol. 12. Kon. Dan. Statist. Bureau, Copenhagen, 1857.

Riel, H. F. van. "Pierre Ancelin." *Tijdschrift voor Kadaster en Landmeetkunde*, 40 (1925), 51–56, 133–44.

Ripley, W. Z. "Notes on Map Making and Graphic Representation." *Quarterly Publications of the American Statistical Association*, 6 (1898), 310–26.

Robinson, Arthur H. "The 1837 Maps of Henry Drury Harness." *Geographical Journal*, 121 (1955), 440–50.

———. "The Cartographic Representation of the Statistical Surface." *International Yearbook of Cartography*, 1 (1961), 53–63.

———. "Mapping the Correspondence of Isarithmic Maps." *Annals of the Association of American Geographers*, 52 (1962), 414–25.

Robinson, A. H., and Bryson, Reid A. "A Method for Describing Quantitatively the Correspondence of Geographical Distributions." *Annals of the Association of American Geographers*, 47 (1957), 379–91.

Robinson, A. H., and Caroe, L. "On the Analysis and Comparison of Statistical Surfaces." *Northwestern University Studies in Geography*, 13 (1967), 252–76.

Robinson, A. H., Lindberg, James B., and Brinkman, Leonard W. "A Correlation and Regression Analysis Applied to Rural Farm Population Densities in the Great Plains." *Annals of the Association of American Geographers*, 51 (1961), 211–21.

Robinson, A. H., and Sale, Randall D. *Elements of Cartography*. 3rd ed. New York: John Wiley, 1969.

Robinson, A. H., and Wallis, Helen M. "Humboldt's Map of Isothermal Lines: A Milestone in Thematic Cartography." *Cartographic Journal*, 4 (1967), 119–23.

Salisbury, Neil E., and LaValle, Placido. "Scale Variations in Morphometric Analysis." Paper delivered at the West Lakes Division Meeting of the Association of American Geographers, Nov. 1–2, 1963.

SELECTED BIBLIOGRAPHY

Schmid, Calvin F. *Handbook of Graphic Presentation.* New York: Ronald Press, 1954.

Schmid, Calvin F., and MacCannell, Earle H. "Basic Problems, Techniques, and Theory of Isopleth Mapping." *Journal of the American Statistical Association,* 50 (1955), 220–39.

Schultz, Gwen M. "An Experiment in Selecting Value Scales for Statistical Distribution Maps." *Surveying and Mapping,* 21 (1961), 224–30.

Shultz, Melvin E., and Richardson, Donald A. *Error Analysis by the Covariance Method.* Aeronautical Chart and Information Center Technical Memorandum no. TM-19, 1962.

Skop, Jacob. "How Accurate Are Our Maps?" *Army,* 10 (1960), 23–25.

Smith, H. Fairfield. "An Experiment Law Describing Heterogeneity in the Yields of Agricultural Crops." *Journal of Agricultural Science,* 28 (1938), 1–23.

Stearns, Franklin. "A Method for Estimating the Quantitative Reliability of Isoline Maps." *Annals of the Association of American Geographers,* 58 (1968), 590–600.

Svaitlovsky, E. E., and Eells, Walter C. "The Centrographical Method and Regional Analysis," *Geographical Review,* 27 (1937), 240–54.

Svensson, Harold. *Method for Exact Characterizing of Denudation Surfaces, Especially Peneplains, as to the Position in Space.* Lund Studies in Geography, Series A, no. 8, 1956.

Tewinkel, G. C. "Analysis of Contour Errors." *Canadian Survey,* 1960, no. 3, pp. 160–65.

Thompson, Morris M. "How Accurate Is That Map?" *Surveying and Mapping,* 16 (1956), 164–73.

Thompson, Morris M., and Davey, Charles H. "Vertical Accuracy of Topographic Maps." *Surveying and Mapping,* 13 (1953), 40–48.

Tobler, Waldo R. "A Polynomial Representation of Michigan Population." *Michigan Academy of Science, Arts, and Letters,* 49 (1964), 445–51.

———. "Automation in the Preparation of Thematic Maps." *Cartographic Journal,* 2 (1965), 32–38.

Uhorczak, Fr. "La Méthode Isarythmique Appliquée Aux Cartes Statistiques." *Polski Przeglad Kartograficzny,* 8 (1930), 95–130.

Wallis, B. C. "Distribution of Nationalities in Hungary." *Geographical Journal,* 47 (1916), 177–88.

Wright, John K. "A Method of Mapping Densities of Population with Cape Cod as an Example." *Geographical Review,* 26 (1936), 103–10.

———. "Map Makers Are Human: Comments on the Subjective in Maps." *Geographical Review,* 32 (1942), 527–44.

———. "A Proposed Atlas of Diseases," Appendix I, "Cartographic Considerations." *Geographical Review,* 34 (1944), 649–52.

———. "The Terminology of Certain Map Symbols." *Geographical Review,* 34 (1944), 653–54.

———. "Crossbreeding Geographical Quantities." *Geographical Review,* 45 (1955), 52–65.

Wright, John K., and others. *Notes on Statistical Mapping with Special Reference to the Mapping of Population Phenomena.* American Geographical Society and Population Association of America, 1938.

Yates, F. "Systematic Sampling." *Philosophical Transactions of the Royal Society of London,* 241 (1948–49), Series A, 345–78.

Zdobnicka, Marja. "Metoda Izarytmiczna W Grafice Statystycznej." *Poklosie Geograficzne,* 1925, pp. 255–71.

Index

Index

Accuracy, map. *See* Map accuracy
Alexander, John W., 11
Analysis of variance: model employed in tests, 53; results of, 53–55; distribution of residuals from ANOVA model, 55
Ancelin, Pierre, 3
ANOVA. *See* Analysis of variance

Blumenstock, David I., 8, 9
Borri, Christoforo, 3
Bruinss, Pieter, 3
Buache, Philippe, 3

Census taking, pattern of enumeration districts for, 71
Center of gravity, for control point location, 9–10
Choropleth mapping, 4
Computer, use of, 3, 10
Control points: location of, 9–10; system of, 9–10; prescription of in isometric mapping, 10; shown on map, 12
Corner, common, 10–11
Correlogram, 29
Cruquius, Nicolas, 3

Dasymetric mapping, 4
"Dead corner," problem of, 10
Declination, mapping of, 3
Discrepancies of resultant maps: derivation of, 49–50; magnitudes of, 55–57; magnitudes in relation to test surfaces, 55–63; magnitudes in relation to test unit area patterns, 56–57; magnitudes in relation to test hexagon sizes, 57; dispersion of, 57–61; dispersion in relation to test surfaces, 61; dispersion in relation to test hexagon sizes, 61; dispersion in relation to test unit area patterns, 61; algebraic values of in relation to test surfaces, 61–63; relation between sign of and sample point location, 62–63
du Carla, Marcellin, 3
Dupain-Triel, Jean Louis, 3
Dupin, Charles, 4

Enumeration districts. *See* Unit areas
Error in isopleth mapping: studies of, 8; due to quality of data, 8–9; due to sampling, 8–9, 12; due to interpolation, 10–12; due to unit area form, 12; variables in, 13. *See also* Discrepancies of resultant maps; Map accuracy
Experimental design of tests, 13

Fidelity of isopleth maps. *See* Discrepancies of resultant maps; Resultant maps
Formula: for geographical mean, 24; for hypsometric quality of European maps, 28; for ANOVA model, 53; for coefficient of variation, 61

Generalization, cartographic: in isarithmic mapping, 6–7; effect of interval on, 7; in isopleth mapping, 9; effect of unit areas on, 12
Generalization in resultant maps: effect of hexagonal transformation on, 27; effect of level of on experiment, 50–52; relation of factors in to sign of resultant map error, 63
Geographical mean, 24

Halley, Edmund, 3
Haxo, François N. B., 3
Hexagonal transformation, usefulness of, 71
Hexagons used in tests: transformation of unit area data to, 24–27; sizes of, 25; in relation to fidelity of resultant maps, 53–55; in relation to magnitudes of discrepancies, 57; optimum size, 57; relation of dispersion of discrepancies to, 61; and relation to visual appearance of fidelity of resultant maps, 66–67, 73
Humboldt, Alexander von, 3

Interpolation: of values from isopleth maps, 5; for location of isopleths, 10; methods of, 10; uncertainty in, 10–11; linear vs. curvilinear, 11–12; triangular arrangement of control points for, 12
Interval, isarithmic, choice of, 8
Isarithm: as generic term, 4; definitions of, 5
Isarithmic technique: early employment of, 3–4; isometric lines in, 4; for volume representation, 4; isopleths in, 4; interval in, 4; conceptual vs. pictorial depiction, 4–5; objectives of, 4–5; commensurability of, 5; interpolation in, 5; generalization in, 6–7; choice of interval, 8; computerization of, 10
Isogonic charts, 3
Isometric mapping: as distinct from isopleth mapping, 4; control points for, 9–10
Isopleth: as distinct from isometric line, 4; smoothing of, 9; interpolation of, 10; effect of shape of unit areas on trend of, 65, 66, 70
Isopleth map: control points for, 5; reading of, 5; interpolation of values from, 5, 11, 12; abstractness of, 5–6; variables in, 6, 73–75; error characteristics of, 8–12; generalization of, 9; control system for, 9–10, 12; gradients on, 10; use of, 64; complexity of, 73–75

Jenks, George F., 8

Kircher, Athanasius, 3

LaValle, Placido, 28

McCarty, Harold H., 64
Mackay, J. Ross, 8, 10
Map accuracy: meaning of, 13; definition of, 14, 61; point vs. area measurement, 28–29; evaluation of, 28–29; volume comparison for, 29; effect of chance location of unit areas on, 50–52; and relation to complexity of surface, 52. *See also* Error in isopleth mapping
Marsigli, Luigi de, 3
Mean, geographical, 24
Meusnier, Jean Baptiste Marie, 3
Mureau, Milet de, 3

National Map Accuracy Standards, 28

Original distributions. *See* Surfaces used in tests

Resultant maps: preparation of, 27; sampling procedure for, 29–30, 49; error near margins of, 30; derivation of test error data from, 49–50; effect of chance location of unit areas on level of generalization in, 50–52; ANOVA tests of, 53–55;

residuals from ANOVA tests of, 55; magnitudes of discrepancies of, 55–57; dispersion of discrepancies of, 57–61; algebraic values of discrepancies of, 61–63
— visual appearance of fidelity of: judgment procedure, 64; in relation to unit area pattern, 64–66, 72–73; in relation to hexagon size, 66–67; in relation to configuration of surfaces, 67–69, 72–73; relative significance of variables in, 69–71

Ripley, W. Z., 8

Salisbury, Neil E., 28, 64
Sampling error, in isopleth mapping, 8–9, 12
Sampling procedure in tests: comparison of possible plans, 29; procedure used in experiment, 29–30, 49; determination of sample size, 30; location of points, 30, 49; derivation of values, 49–50
Statistical surface, 4–5, 10
Surfaces used in tests: complexity of, 14; selection of for experiment, 14, 19; relation to magnitude of discrepancies, 55–63; relation to dispersion of discrepancies, 61; relation to algebraic discrepancies, 61–63; relation to visual appearance of fidelity of resultant maps, 67–69, 72–73

Transformation, hexagonal, usefulness of, 71

Transformation of unit area data: procedure, 24–27; disadvantage of, 27

Uhorczak, Fr., 10, 11, 28
Unit areas: patterns of, 12; transformation of, 12
Unit areas employed in the tests: selection of for testing, 19; size and variance of, 19; derivation of experimental data from, 19, 24; transformation of data to hexagonal pattern, 24–27; effect of chance location on generalization, 50–52; relation of size and shape to fidelity of resultant maps, 53–55; pattern in relation to magnitude of discrepancies, 56–57; relation of dispersion of discrepancies to patterns of, 61; relation to visual appearance of the fidelity of resultant maps, 64–66, 72–73; effect of shape on trend of isopleths, 65, 66, 70

Variance, analysis of. *See* Analysis of variance
Visual analysis. *See* Resultant maps, visual appearance of fidelity of

Wright, John K., 8, 12

Zahorchak, George A., 11
Zdobnicka, M., 28